2995

The Checkbook Series

Electrical and Electronic Principles 3 Checkbook

Second edition

J O Bird
BSc(Hons), CEng MIEE, FcollP, FIMA, MIElecIE

A J C May
BA, CEng, MIMechE, FIElecIE, MBIM

Newnes
An imprint of Butterworth-Heinemann Ltd
Linacre House, Jordan Hill, Oxford OX2 8DP

A member of the Reed Elsevier group

OXFORD LONDON BOSTON
MUNICH NEW DELHI SINGAPORE SYDNEY
TOKYO TORONTO WELLINGTON

First published 1981
Reprinted 1986
Second edition 1989
Reprinted 1991, 1992, 1993

British Library Cataloguing in Publication Data
Bird, J. O.
Electrical engineering
1. Questions and answers for technicians
I. Title II. May A. J. C.
621.3076

ISBN 0 7506 0336 4

Printed and bound in Great Britain by
Redwood Press Limited, Melksham, Wiltshire

Contents

Note to readers

Checkbooks are designed for students seeking technician or equivalent qualification through the courses of the Business and Technician Education Council (BTEC), the Scottish Technical Education Council, Australian Technical and Further Education Departments, East and West African Examinations Council and other comparable examining authorities in technical subjects.

Checkbooks use problems and worked examples to establish and exemplify the theory contained in technical syllabuses. *Checkbook* readers gain real understanding through seeing problems solved and through solving problems themselves. *Checkbooks* do not supplant fuller textbooks, but rather supplement them with an alternative emphasis and an ample provision of worked and unworked problems, essential data, short answers and multi-choice questions (with answers where possible).

Preface

This textbook of worked problems provides coverage of selected material from the Business and Technician Education Council's Bank of objectives in Electrical and Electronic Principles at NIII level. However it can also be regarded as a basic textbook in Electrical Principles for a much wider range of courses. It provides a follow-up to the *Electrical and Electronic Principles 2 Checkbook*.

The aim of the book is to introduce students to the basic electrical principles needed by technicians in fields such as electrical engineering, electronics and telecommunications areas.

Each topic considered in the text is presented in a way that assumes in the reader only the knowledge attained at BTEC level II (or equivalent) in Electrical and Electronic Principles and in Mathematics.

This practical second edition electrical and electronic principles book contains some 200 illustrations, nearly 150 detailed worked problems, followed by some 450 further problems with answers.

This second edition of the book incorporates new material on Nortons theorem, modulation, filter and attenuation circuits, in addition to a number of other minor amendments.

The authors would like to express their appreciation for the friendly co-operation and helpful advice given to them by the publishers.

J O Bird
A J C May
Highbury College of Technology
Portsmouth

1 Circuit theorems

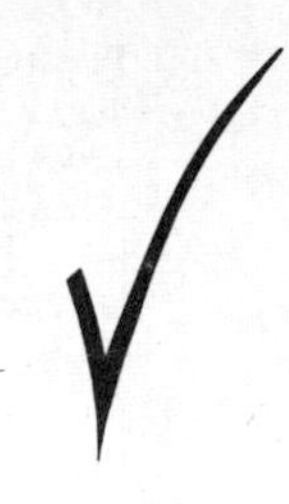

A. MAIN POINTS CONCERNED WITH D.C. CIRCUIT ANALYSIS

1 The laws which determine the currents and voltage drops in d.c. networks are: (a) Ohm's law, (b) the laws for resistors in series and in parallel, and (c) Kirchhoff's laws. In addition, there are a number of circuit theorems which have been developed for solving problems in electrical networks. These include:
   (i) the superposition theorem,
   (ii) Thévenin's theorem,
   (iii) Norton's theorem, and
   (iv) the maximum power transfer theorem.

2 The **superposition theorem** states:
'In any network made up of linear resistances and containing more than one source of emf, the resultant current flowing in any branch is the algebraic sum of the currents that would flow in that branch if each source was considered separately, all other sources being replaced at that time by their respective internal resistances.'
(See *Problems 1 and 2*)

3 The following points involving d.c. circuit analysis need to be appreciated before proceeding with problems using Thévenin's and Norton's theorems:
   (i) The open–circuit voltage, E, across terminals AB in *Fig 1* is equal to 10 V, since no current flows through the 2 Ω resistor and hence no voltage drop occurs.
   (ii) The open-circuit voltage, E, across terminals AB in *Fig 2(a)* is the same as the voltage across the 6 Ω resistor. The ciruit may be redrawn as shown in *Fig 2(b)*

$$E = \left(\frac{6}{6+4}\right)(50)$$

by voltage division in a series circuit,
i.e. E = **30 V**

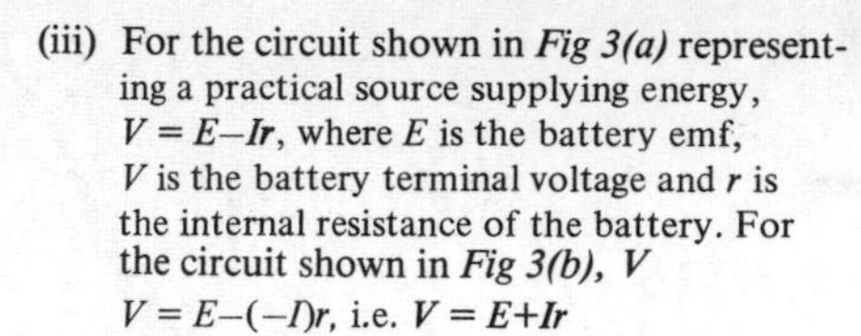
(iii) For the circuit shown in *Fig 3(a)* representing a practical source supplying energy, $V = E - Ir$, where E is the battery emf, V is the battery terminal voltage and r is the internal resistance of the battery. For the circuit shown in *Fig 3(b)*, V
$V = E - (-I)r$, i.e. $V = E + Ir$

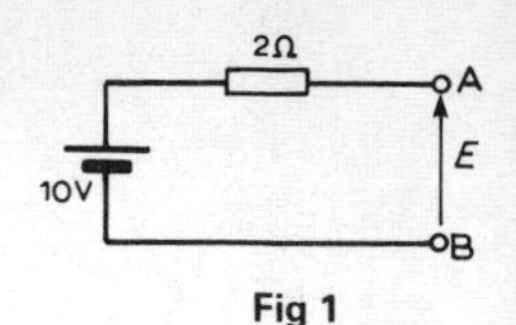

Fig 1

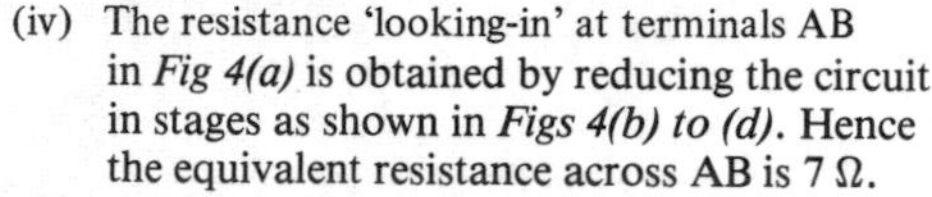
(iv) The resistance 'looking-in' at terminals AB in *Fig 4(a)* is obtained by reducing the circuit in stages as shown in *Figs 4(b) to (d)*. Hence the equivalent resistance across AB is 7 Ω.

(v) For the circuit shown in *Fig 5(a)*, the 3 Ω resistor carries no current and the p.d. across the 20 Ω resistor is 10 V. Redrawing the circuit gives *Fig 5(b)*, from which

$$E = \left(\frac{4}{4+6}\right) \times 10 = \mathbf{4\ V}$$

(vi) If the 10 V battery in *Fig 5(a)* is removed and replaced by a short-circuit, as shown in *Fig 5(c)*, then the 20 Ω resistor may be removed. The reason for this is that a short-circuit has zero resistance, and 20 Ω in parallel with zero ohms gives an equivalent

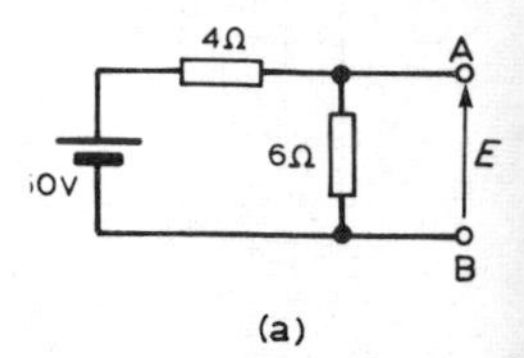

(a)

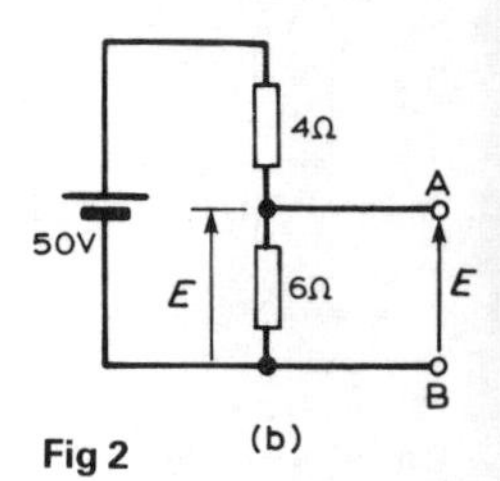

Fig 2 (b)

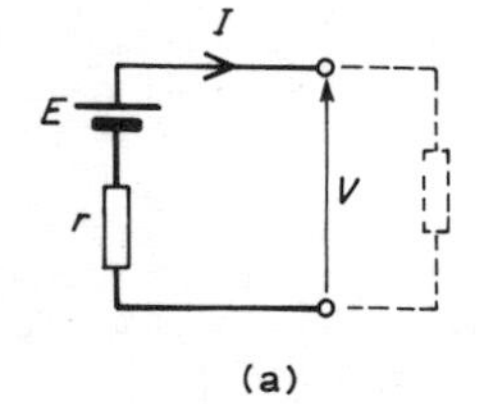

(a)

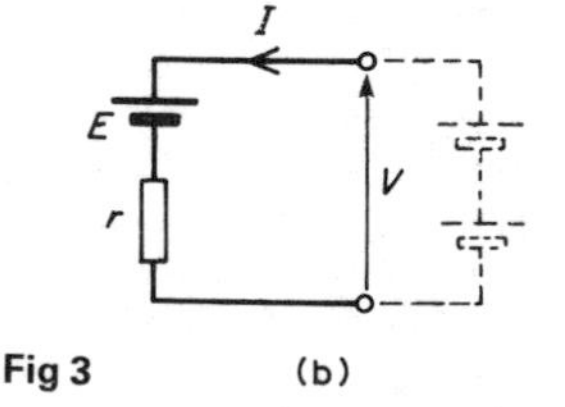

(b)

Fig 3

resistance of $\frac{20 \times 0}{20 + 0}$, i.e. 0 Ω. The circuit is then as shown in *Fig 5(d)*, which is redrawn in *Fig 5(e)*. From *Fig 5(e)*, the equivalent resistance across AB.

$$r = \frac{6 \times 4}{6 + 4} + 3 = 2.4 + 3 = \mathbf{5.4\ \Omega}$$

(vii) To find the voltage across AB in *Fig 6*:
Since the 20 V supply is across the 5 Ω and 15 Ω resistors in series then, by voltage division, the voltage drop across AC, $V_{AC} = \left(\frac{5}{5+15}\right)(20) = 5$ V.

Similarly, $V_{CB} = \left(\frac{12}{12+3}\right)(20) = 16$ V. V_C is at a potential of +20 V.

$V_A = V_C - V_{AC} = +20-5 = 15$ V and $V_B = V_C - V_{BC} = +20-16 = 4$ V.
Hence the voltage between AB is $V_A - V_B = 15-4 = 11$ V and current would flow from A to B since A has a higher potential than B.

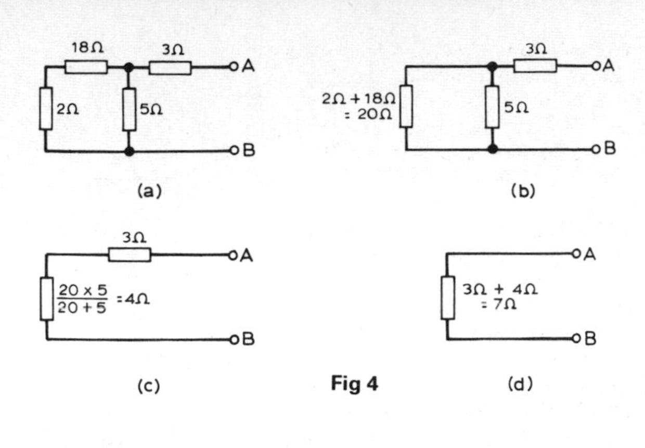

Fig 4

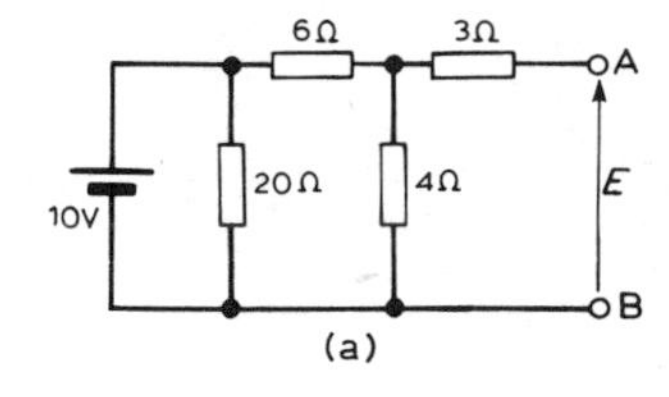

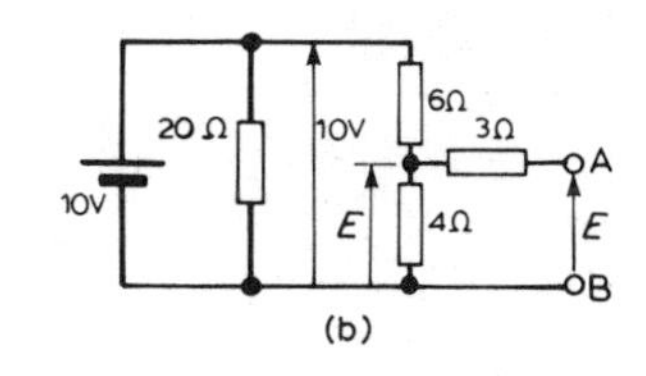

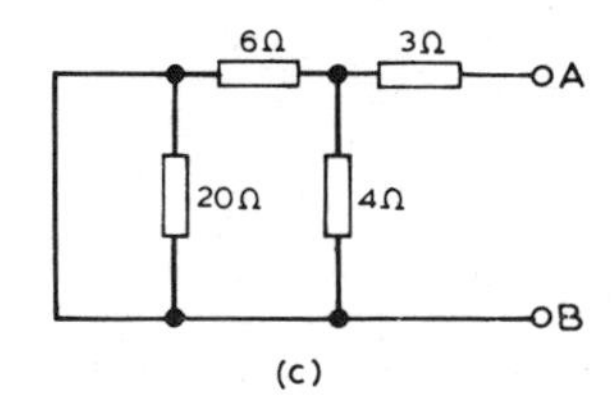

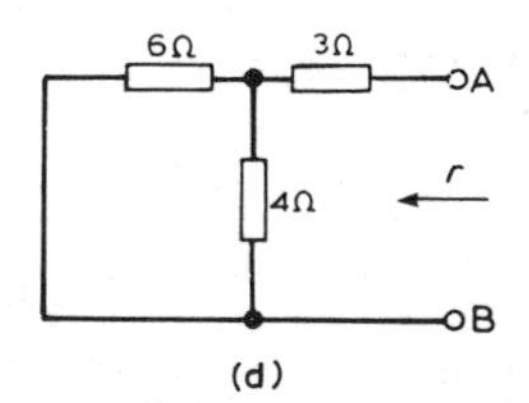

Fig 5

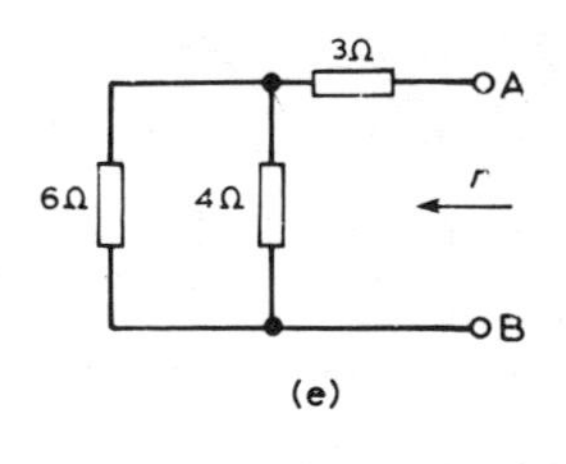

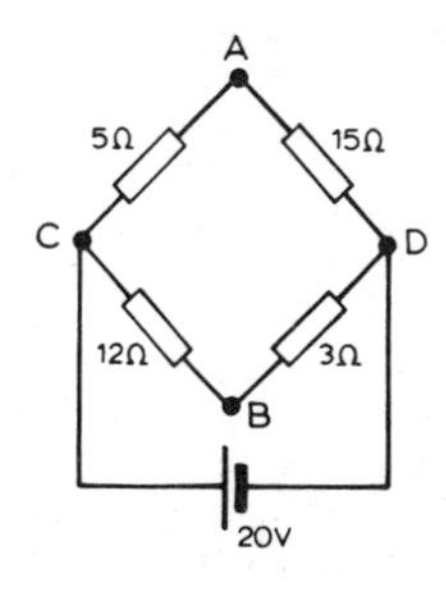

Fig 6

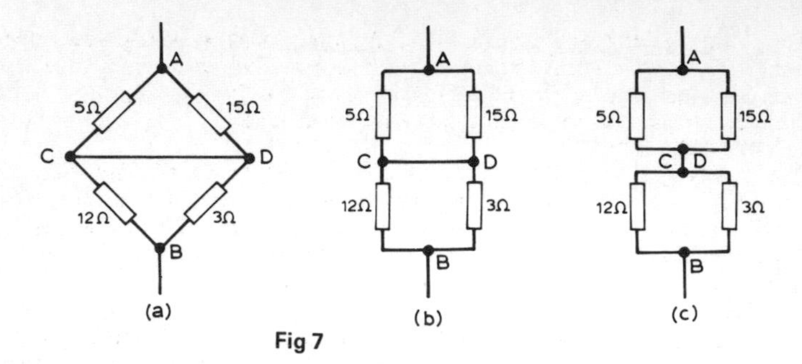

Fig 7

(viii) In *Fig 7(a)*, to find the equivalent resistance across AB the circuit may be redrawn as in *Figs 7(b) and (c)*. From *Fig 7(c)*, the equivalent resistance across AB

$$= \frac{5 \times 15}{5 + 15} + \frac{12 \times 3}{12 + 3}$$

$= 3.75 + 2.4 =$ **6.15 Ω**

(ix) In the worked problems in section B it may be considered that Thévenin's and Norton's theorems have no obvious advantages compared with, say, Kirchhoff's laws. However, these theorems can be used to analyse part of a circuit and in much more complicated networks the principle of replacing the supply by a constant voltage source in series with a resistance (or impedance) is very useful.

4 **Thévenin's theorem** states:
'The current in any branch of a network is that which would result if an emf, equal to the p.d. across a break made in the branch, were introduced into the branch, all other emf's being removed and represented by the internal resistances of the sources.'

5 The procedure adopted when using Thévenin's theorem is summarized below. To determine the current in any branch of an active network (i.e. one containing a source of emf):

(i) remove the resistance R from that branch,
(ii) determine the open-circuit voltage, E, across the break,
(iii) remove each source of emf and replace them by their internal resistances and then determine the resistance, r, 'looking-in' at the break,
(iv) determine the value of the current from the equivalent circuit shown in *Fig 8*, i.e. $I = \frac{E}{R + r}$
(See *Problems 3 to 8*).

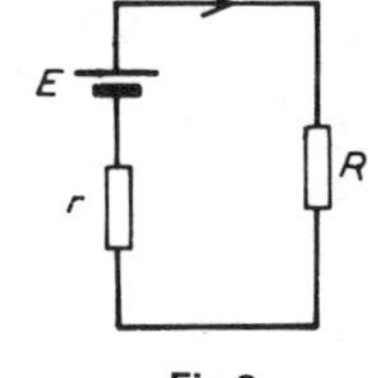

Fig 8

6 A source of electrical energy can be represented by a source of emf in series with a resistance. In para. 5, the Thévenin constant–voltage source consisted of a constant emf E in series with an internal resistance r. However this is not the only form of representation. A source of electrical energy can also be represented by a constant–current source in parallel with a resistance. It may be

shown that the two forms are equivalent. An **ideal constant–voltage generator** is one with zero internal resistance so that it supplies the same voltage to all loads. An **ideal constant–current generator** is one with infinite internal resistance so that it supplies the same current to all loads.

7 **Norton's theorem** states:
'The current that flows in any branch of a network is the same as that which would flow in the branch if it were connected across a source of electrical energy, the short–circuit current of which is equal to the current that would flow in a short–circuit across the branch, and the internal resistance of which is equal to the resistance which appears across the open–circuited branch terminals.'

8 The procedure adopted when using Norton's theorem is summarized below.
To determine the current flowing in a resistance R of a branch AB of an active network:
(i) short–circuit branch AB
(ii) determine the short–circuit current I_{sc} flowing in the branch
(iii) remove all sources of emf and replace them by their internal resistance (or, if a current source exists, replace with an open–circuit), then determine the resistance r, 'looking-in' at a break made between A and B
(iv) determine the current I flowing in resistance R from the Norton equivalent network shown in *Fig 9*, i.e.

$$I = \left(\frac{r}{r+R}\right) I_{sc}$$

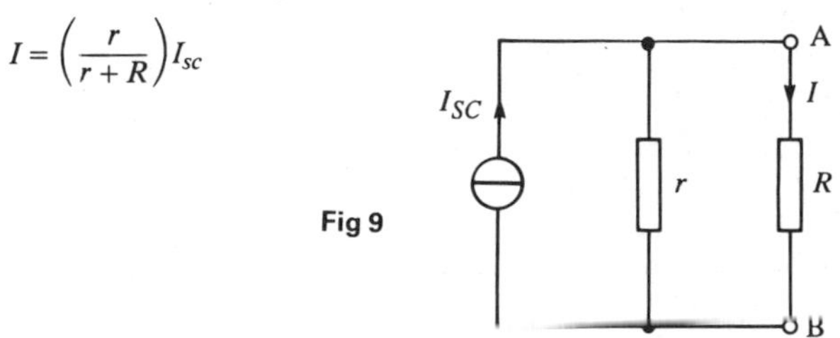

Fig 9

Note the symbol for an ideal current source (BS 3939, 1985) shown in *Fig 9* (See *Problems 9 to 13*)

9 The Thévenin and Norton networks shown in *Fig 10* are equivalent to each other. The resistance 'looking-in' at terminals AB is the same in each of the networks, i.e. r.

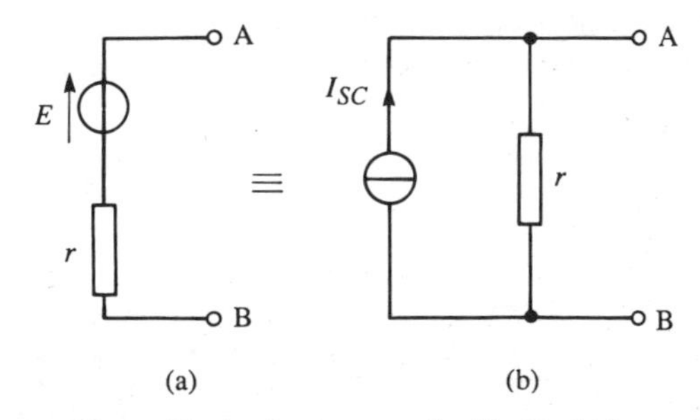

Fig 10

(Note the symbol for an ideal voltage source in *Fig 10* which may be used as an alternative to the battery symbol.)

If terminals AB in *Fig 10(a)* are short–circuited, the short–circuit current is given by $\frac{E}{r}$. If terminals AB in *Fig 10(b)* are short–circuited, the short–circuit current is I_{sc}. For the circuit shown in *Fig 10(a)* to be equivalent to the circuit in *Fig 10(b)* the same short–circuit current must flow. Thus $I_{sc} = \frac{E}{r}$. *Fig 11* shows a source of emf E in series with a resistance r feeding a load resistance R.

From *Fig 11*, $I = \frac{E}{r+R} = \frac{E/r}{(r+R)/r} = \left(\frac{r}{r+R}\right)\frac{E}{r}$

$$\text{i.e. } I = \left(\frac{r}{r+R}\right) I_{sc}$$

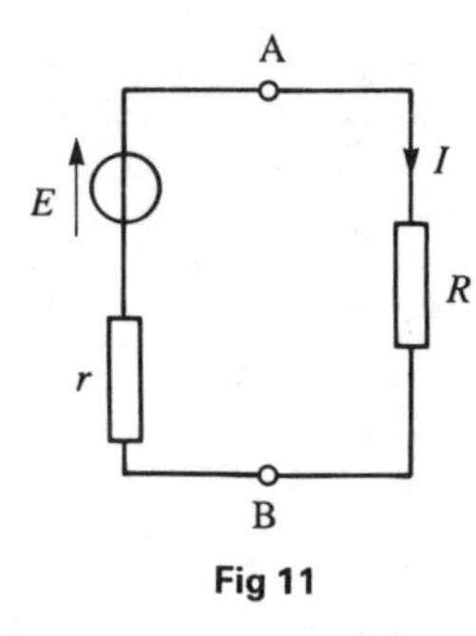

Fig 11

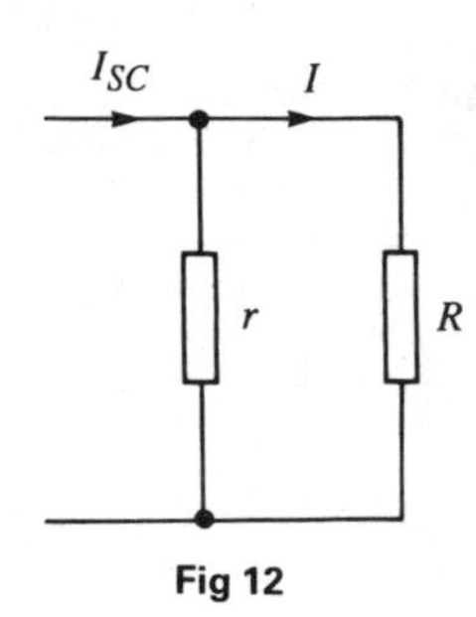

Fig 12

From *Fig 12*, it can be seen that, when viewed from the load, the source appears as a source of current I_{sc} which is divided between r and R connected in parallel.

Thus it is shown that the two representations shown in *Fig 10* are equivalent. (See *Problems 14 to 17*)

10 The **maximum power transfer theorem** states:
'The power transferred from a supply source to a load is at its maximum when the resistance of the load is equal to the internal resistance of the source'.
Hence, in *Fig 13*, when $R = r$ the power transferred from the source to the load is a maximum.

11 Varying a load resistance to be equal, or almost equal, to the source internal resistance is called matching. Examples where resistance matching is important include coupling an aerial to a transmitter or receiver, or in coupling a loudspeaker to an amplifier where coupling transformers may be used to give maximum power transfer.

With d.c. generators or secondary cells, the internal resistance is usually very small. In such cases, if an attempt is made to make the load resistance as small as the source internal resistance, overloading of the source results.

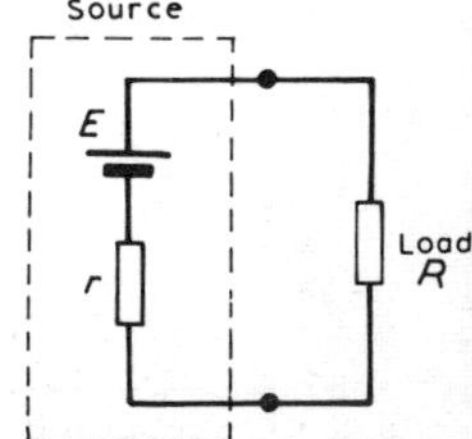

Fig 13

12 A method of achieving maximum power transfer between a source and a load is to adjust the value of the load resistance to 'match' the source internal resistance. A transformer may be used as a **resistance matching device** by connecting

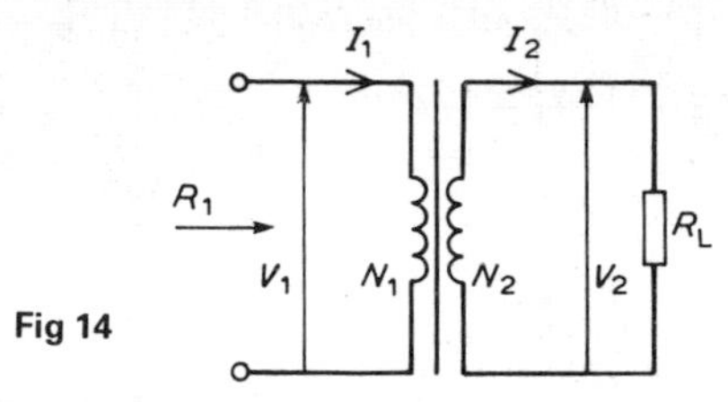

Fig 14

it between the load and the source. The reason why a transformer can be used for this is shown below. With reference to *Fig 14*:

$$R_L = \frac{V_2}{I_2} \text{ and } R_1 = \frac{V_1}{I_1}$$

For an ideal transformer, $V_1 = \left(\frac{N_1}{N_2}\right) V_2$ and $I_1 = \left(\frac{N_2}{N_1}\right) I_2$

Thus the equivalent input resistance R_1 of the transformer is given by:

$$R_1 = \frac{V_1}{I_1} = \frac{\left(\frac{N_1}{N_2}\right) V_2}{\left(\frac{N_2}{N_1}\right) I_2} = \left(\frac{N_1}{N_2}\right)^2 \frac{V_2}{I_2} = \left(\frac{N_1}{N_2}\right)^2 R_L$$

Hence by varying the value of the turns ratio, the equivalent input resistance of a transformer can be 'matched' to the internal resistance of a load to achieve maximum power transfer.
(See *Problems 18 to 25*)

B. WORKED PROBLEMS ON CIRCUIT THEOREMS

SUPERPOSITION THEOREM

Problem 1 Fig 15 shows a circuit containing two sources of emf, each with their internal resistance. Determine the current in each branch of the network by using the superposition theorem.

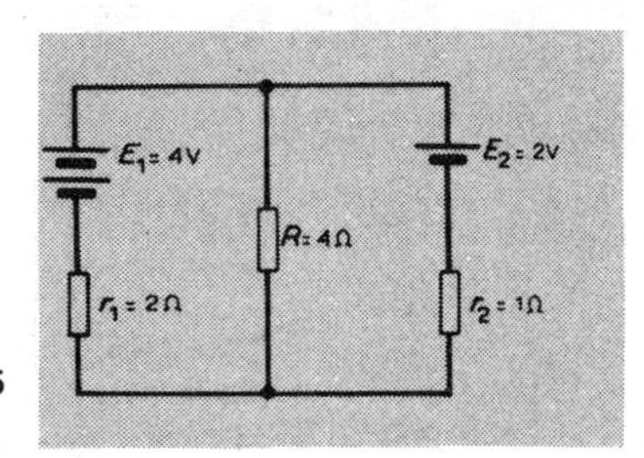

Fig 15

Procedure:

1 Redraw the original circuit with source E_2 removed, being replaced by r_2 only, as shown in *Fig 16(a)*.

2 Label the currents in each branch and their directions as shown in *Fig 16(a)* and

determine their values. (Note that the choice of current directions depends on the battery polarity, which, by convention is taken as flowing from the positive battery terminal as shown.) R in parallel with r_2 gives an equivalent resistance of

$$\frac{4 \times 1}{4+1} = 0.8\ \Omega$$

From the equivalent circuit of *Fig 16(b)*

$$I_1 = \frac{E_1}{r_1 + 0.8} = \frac{4}{2+0.8}$$

$$= 1.429\ \text{A}$$

From *Fig 16(a)*

$$I_2 = \left(\frac{1}{4+1}\right) I_1 = \frac{1}{5}(1.429) = 0.286\ \text{A}$$

and

$$I_3 = \left(\frac{4}{4+1}\right) I_1 = \frac{4}{5}(1.429) = 1.143\ \text{A}$$

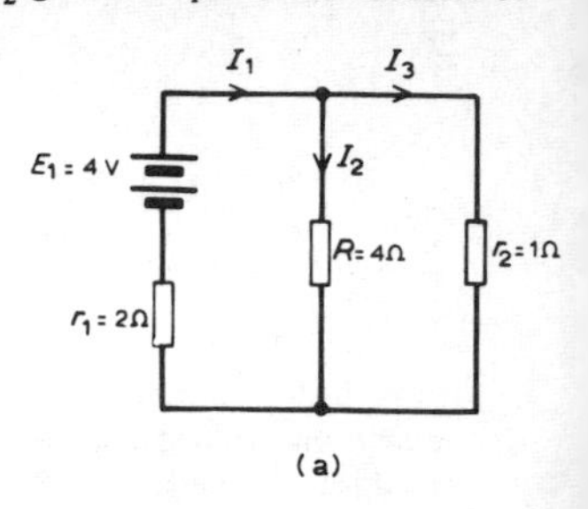

(a)

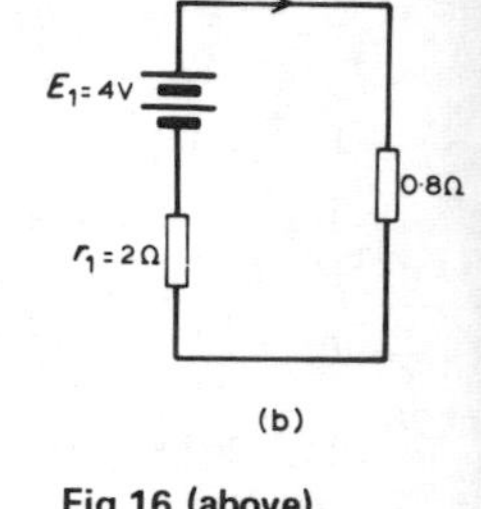

(b)

Fig 16 (above)

3 Redraw the original circuit with source E_1 removed, being replaced by r_1 only, as shown in *Fig 17(a)*.

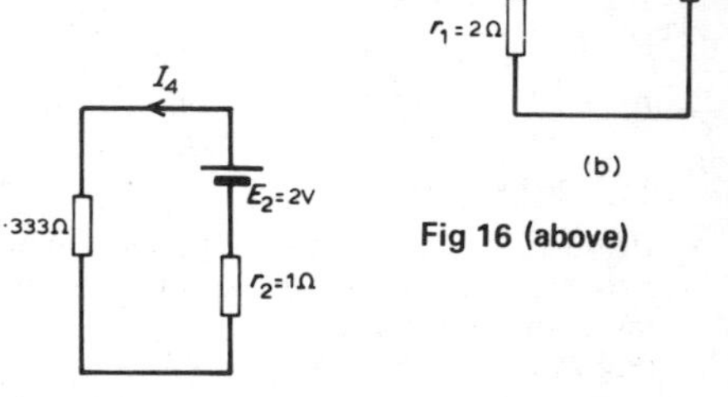

(a) (b)

Fig 17 (left)

4 Label the currents in each branch and their directions as shown in *Fig 17(a)* and determine their values. r_1 in parallel with R gives an equivalent resistance of

$$\frac{2 \times 4}{2+4} = \frac{8}{6} = 1.333\ \Omega$$

From the equivalent circuit of *Fig 17(b)*

$$I_4 = \frac{E_2}{1.333 + r_2} = \frac{2}{1.333+1}$$

$$= 0.857\ \text{A}$$

From *Fig 17(a)*

$$I_5 = \left(\frac{2}{2+4}\right) I_4 = \frac{2}{6}(0.857) = 0.286\ \text{A}$$

$$I_6 = \left(\frac{4}{2+4}\right) I_4 = \frac{4}{6}(0.857) = 0.571\ \text{A}$$

5 Superimpose *Fig 17(a)* on to *Fig 16(a)* as shown in *Fig 18*.

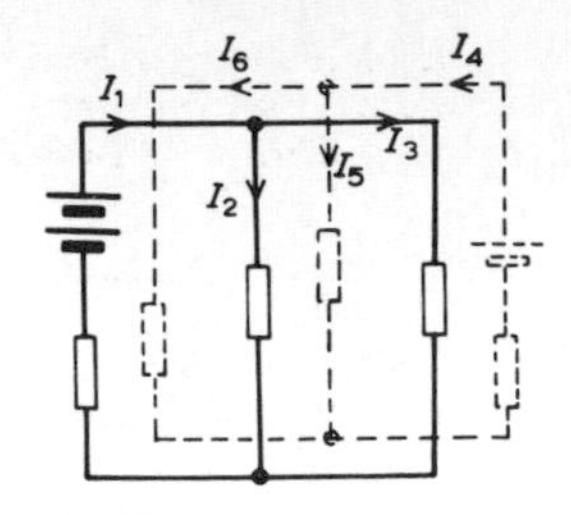

Fig 18

Fig 19

6 Determine the algebraic sum of the currents flowing in each branch. Resultant current flowing through source 1, i.e.

$$I_1 - I_6 = 1.429 - 0.571$$
$$= \mathbf{0.858\ A\ (discharging)}$$

Resultant current flowing through source 2, i.e.

$$I_4 - I_3 = 0.857 - 1.143$$
$$= \mathbf{-0.286\ A\ (charging)}$$

Resultant current flowing through resistor R, i.e.

$$I_2 + I_5 = 0.286 + 0.286$$
$$= \mathbf{0.572\ A}$$

The resultant currents with their directions are shown in *Fig 19.*

Problem 2 For the circuit shown in *Fig 20*, find, using the superposition theorem, (a) the current flowing in and the pd across the 18 Ω resistor, (b) the current in the 8 V battery and (c) the current in the 3 V battery.

1 Removing source E_2 gives the circuit of *Fig 21(a).*
2 The current directions are labelled as shown in *Fig 21(a)*, I_1 flowing from the positive terminal of E_1.
From *Fig 21(b)*

$$I_1 = \frac{E_1}{3+1.8} = \frac{8}{4.8} = 1.667\ \text{A}$$

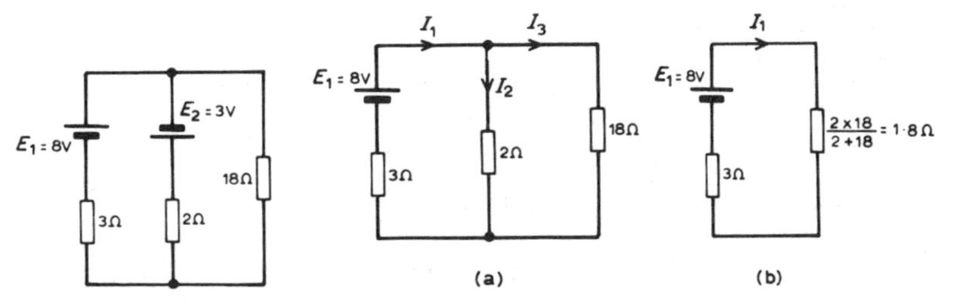

Fig 20

Fig 21

From *Fig 21(a)*

$$I_2 = \left(\frac{18}{2+18}\right) I_1 = \frac{18}{20}\ (1.667) = 1.500\ \text{A}$$

and

$$I_3 = \left(\frac{2}{2+18}\right) I_1 = \frac{2}{20}\ (1.667) = 0.167\ \text{A}$$

3 Removing source E_1 gives the circuit of *Fig 22(a)* (which is the same as *Fig 22(b)*).

4 The current directions are labelled as shown in *Figs 22(a) and 22(b)*, I_4 flowing from the positive terminal of E_2.

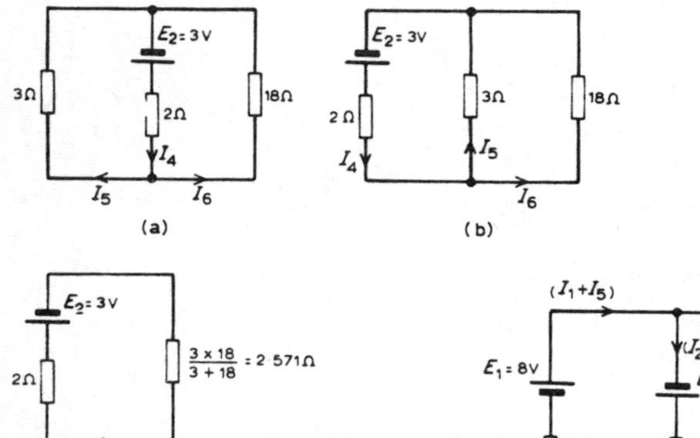

Fig 22 (left)

Fig 23 (below)

From *Fig 22(c)*

$$I_4 = \frac{E_2}{2+2.571} = \frac{3}{4.571} = 0.656\ \text{A}$$

From *Fig 22(b)*

$$I_5 = \left(\frac{18}{3+18}\right) I_4$$

$$= \frac{18}{21}\ (0.656) = 0.562\ \text{A}$$

$$I_6 = \left(\frac{3}{3+18}\right) I_4$$

$$= \frac{3}{21}\ (0.656) = 0.094\ \text{A}$$

5 Superimposing *Fig 22(a)* on to *Fig 21(a)* gives the circuit in *Fig 23*.

6 (a) Resultant current in the 18 Ω resistor = $I_3 - I_6$ = 0.167–0.094 = 0.073 A
Pd across the 18 Ω resistor = 0.073 × 18 = **1.314 V**

(b) Resultant current in the 8 V battery = $I_1 + I_5$ = 1.667+0.562 = **2.229 A (discharging)**

(c) Resultant current in the 3 V battery $I_2 + I_4$ = 1.500+0.656 = **2.156 A (discharging)**

Problem 3 Use Thévenin's theorem to find the current flowing in the 10 Ω resistor for the circuit shown in *Fig 24(a)*.

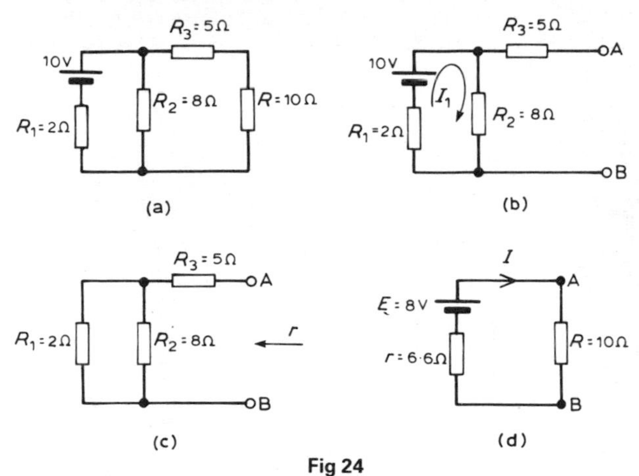

Fig 24

Following the procedure in para 5:

(i) The 10 Ω resistance is removed from the circuit as shown in *Fig 24(b)*.

(ii) There is no current flowing in the 5 Ω resistor and current I_1 is given by

$$I_1 = \frac{10}{R_1+R_2} = \frac{10}{2+8} = 1 \text{ A.}$$

P.d. across $R_2 = I_1 R_2 = 1 \times 8 = 8$ V.

Hence p.d. across AB, i.e. the open-circuit voltage across the break, $E = 8$ V.

(iii) Removing the source of emf gives the circuit of *Fig 24(c)*.

$$\text{Resistance, } r = R_3 + \frac{R_1 R_2}{R_1+R_2} = 5 + \frac{2 \times 8}{2+8} = 5+1.6 = 6.6\ \Omega$$

(iv) The equivalent Thévenin's circuit is shown in *Fig 24(d)*.

$$\text{Current } I = \frac{E}{R+r} = \frac{8}{10+6.6} = \frac{8}{16.6} = 0.482 \text{ A}$$

Hence the current flowing in the 10 Ω resistor of *Fig 24(a)* is 0.482 A

Problem 4 For the network shown in *Fig 25(a)* determine the current in the 0.8 Ω resistor using Thévenin's theorem.

Following the procedure of para 5:

(i) The 0.8 Ω resistor is removed from the circuit as shown in *Fig 25(b)*.

(ii) $\text{Current } I_1 = \dfrac{12}{1+5+4} = \dfrac{12}{10} = 1.2 \text{ A}$

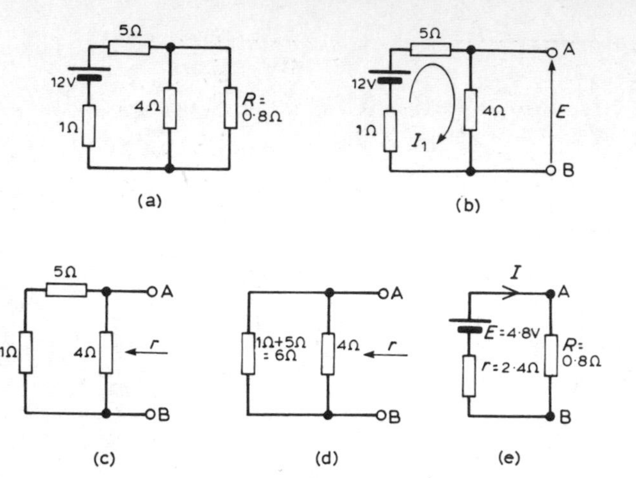

Fig 25

P.d. across 4 Ω resistor = $4I_1$ = (4)(1.2) = 4.8 V
Hence p.d. across AB, i.e. the open-circuit voltage across AB, E = 4.8 V

(iii) Removing the source of emf gives the circuit shown in *Fig 25(c)*. The equivalent circuit of *Fig 25(c)* is shown in *Fig 25(d)*, from which,

$$\text{resistance } r = \frac{4 \times 6}{4+6} = \frac{24}{10} = 2.4\ \Omega.$$

(iv) The equivalent Thévenin's circuit is shown in *Fig 25(e)*, from which,

$$\text{current } I = \frac{E}{r+R} = \frac{4.8}{2.4+0.8} = \frac{4.8}{3.2}$$

$$= \mathbf{1.5A = current\ in\ the\ 0.8\ \Omega\ resistor.}$$

Problem 5 Use Thévenin's theorem to determine the current I flowing in the 4 Ω resistor shown in *Fig 26(a)*. Find also the power dissipated in the 4 Ω resistor.

Following the procedure of para 5:

(i) The 4 Ω resistor is removed from the circuit as shown in *Fig 26(b)*.

(ii) Current $I_1 = \dfrac{E_1 - E_2}{r_1 + r_2} = \dfrac{4-2}{2+1} = \dfrac{2}{3}$ A.

P.d. across AB, $E = E_1 - I_1 r_1 = 4 - \frac{2}{3}(2) = 2\frac{2}{3}$ V (see para 3(iii)).

(Alternatively, p.d. across AB, $E = E_2 - I_1 r_2 = 2 - \left(-\frac{2}{3}\right)(1) = 2\frac{2}{3}$ V.

(iii) Removing the sources of emf gives the circuit shown in *Fig 26(c)*, from which resistance $r = \dfrac{2 \times 1}{2+1} = \frac{2}{3}\ \Omega$.

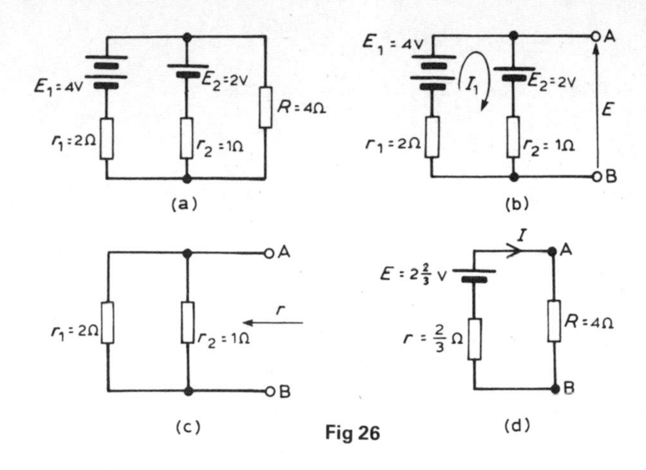

Fig 26

(iv) The equivalent Thévenin's circuit is shown in *Fig 26(d)*, from which

$$\text{current, } I = \frac{E}{r+R} = \frac{2\frac{2}{3}}{\frac{2}{3}+4} = \frac{8/3}{14/3} = \frac{8}{14} = \mathbf{0.571\ A}$$

$= \textbf{current in the 4 } \boldsymbol{\Omega} \textbf{ resistor.}$

Power dissipated in 4 Ω resistor, $P = I^2R = (0.571)^2(4) = \mathbf{1.304\ W}$

Problem 6 Determine the current in the 5 Ω resistance of the network shown in *Fig 27(a)* using Thévenin's theorem. Hence find the currents flowing in the other two branches.

Following the procedure of para 5:

(i) The 5 Ω resistance is removed from the circuit as shown in *Fig 27(b)*.

(ii) Current $I_1 = \frac{12+4}{0.5+2} = \frac{16}{2.5} = 6.4$ A

P.d. across AB, $E = E_1 - I_1r_1 = 4-(6.4)(0.5) = 0.8$ V (see para 3(iii)).
(Alternatively, $E = E_2 - I_1r_1 = -12-(-6.4)(2) = 0.8$ V)

(iii) Removing the sources of emf gives the circuit shown in *Fig 27(c)*, from which

$$\text{resistance } r = \frac{0.5 \times 2}{0.5+2} = \frac{1}{2.5} = 0.4\ \Omega$$

(iv) The equivalent Thévenin's circuit is shown in *Fig 27(d)*, from which,

$$\text{current } I = \frac{E}{r+R} = \frac{0.8}{0.4+5} = \frac{0.8}{5.4} = \mathbf{0.148\ A}$$

$= \textbf{current in the 5 } \boldsymbol{\Omega} \textbf{ resistor}$

From *Fig 27(e)*, voltage $V = IR_3 = (0.148)(5) = 0.74$ V.

From para 3(iii), $V = E_1 - I_Ar_1$
i.e. $0.74 = 4-(I_A)(0.5)$

Hence current, $I_A = \frac{4-0.74}{0.5} = \frac{3.26}{0.5} = \mathbf{6.52\ A}$

Also from *Fig 27(e)*, $V = -E_2 + I_B r_2$ (see para 3(iii))

i.e. $0.74 = -12+(I_B)(2)$

Hence current $I_B = \frac{12+0.74}{2} = \frac{12.74}{2} = \mathbf{6.37\ A}$

[Check, from *Fig 27(e)*, $I_A = I_B + I$, correct to 2 significant figures (Kirchhoff's current law).]

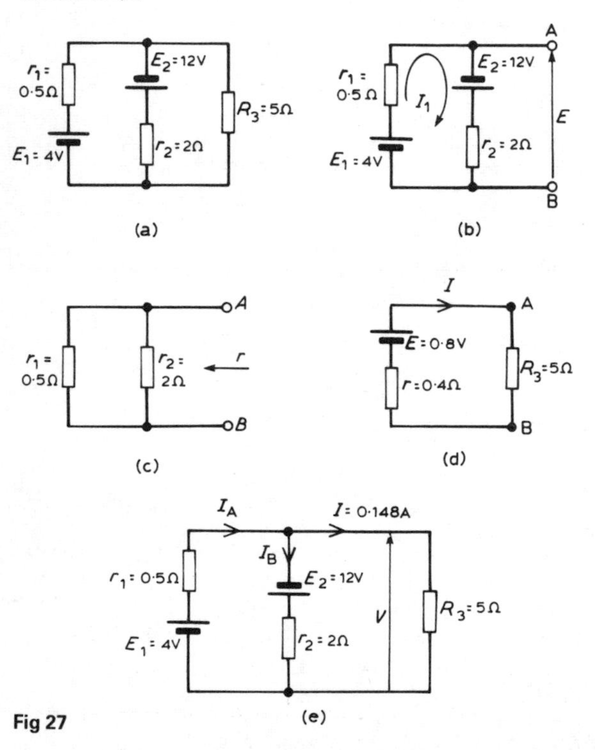

Fig 27

Problem 7 Use Thévenin's theorem to determine the current flowing in the 3 Ω resistance of the network shown in *Fig 28(a)*. The voltage source has negligible internal resistance.

Following the procedure of para 5:

(i) The 3 Ω resistance is removed from the circuit as shown in *Fig 28(b)*.

(ii) The $1\frac{2}{3}$ Ω resistance now carries no current.

P.d. across 10 Ω resistor $= \left(\frac{10}{10+5}\right)$ (24) = 16 V (see para 3(v)).

Hence p.d. across AB, E = 16 V

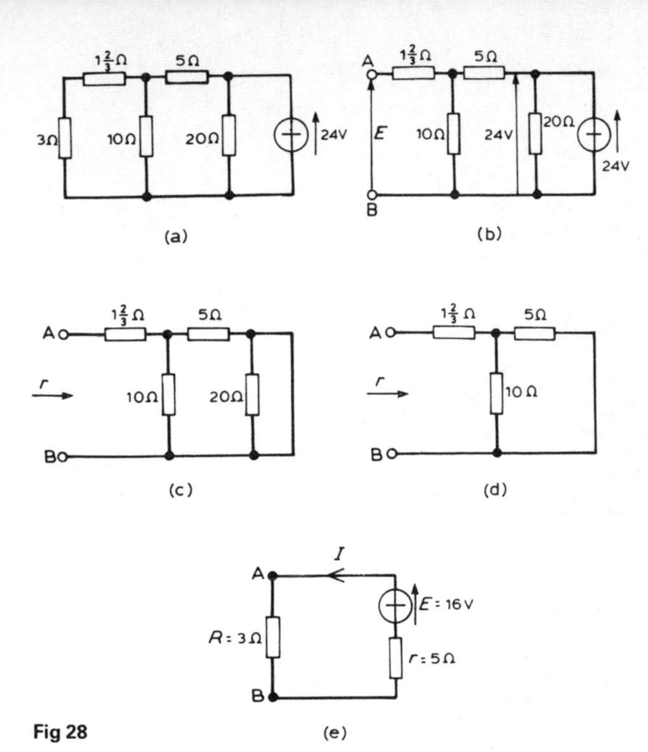

Fig 28

(iii) Removing the source of emf and replacing it by its internal resistance means that the 20 Ω resistance is short-circuited as shown in *Fig 28(c)* since its internal resistance is zero. The 20 Ω resistor may thus be removed as shown in *Fig 28(d)* (see para 3(vi)).

From *Fig 28(d)*, resistance, $r = 1\frac{2}{3} + \frac{10 \times 5}{10+5} = 1\frac{2}{3} + \frac{50}{15} = 5\ \Omega$

(iv) The equivalent Thévenin's circuit is shown in *Fig 28(e)*, from which

current, $I = \frac{E}{r+R} = \frac{16}{3+5} = \frac{16}{8} = 2$ A = current in the 3 Ω resistance

Problem 8 A Wheatstone Bridge network is shown in *Fig 29(a)*. Calculate the current flowing in the 32 Ω resistor, and its direction, using Thévenin's theorem. Assume the source of emf to have negligible resistance.

Following the procedure of para 5:

(i) The 32 Ω resistor is removed from the circuit as shown in *Fig 29(b)*.

(ii) The p.d. between A and C, $V_{AC} = \left(\frac{R_1}{R_1+R_4}\right)(E) = \left(\frac{2}{2+11}\right)(54)$

$= 8.31$ V

The p.d. between B and C, $V_{BC} = \left(\frac{R_2}{R_2+R_3}\right)(E) = \left(\frac{14}{14+3}\right)(54)$

$= 44.47 \text{ V}$

Hence the p.d. between A and B = 44.47–8.31 = 36.16 V
Point C is at a potential of +54 V. Between C and A is a voltage drop of 8.31 V. Hence the voltage at point A is 54–8.31 = 45.69 V. Between C and B is a voltage drop of 44.47 V. Hence the voltage at point B is 54–44.47 = 9.53 V. Since the voltage at A is greater than at B, current must flow in the direction A to B. (See para 3(vii)).

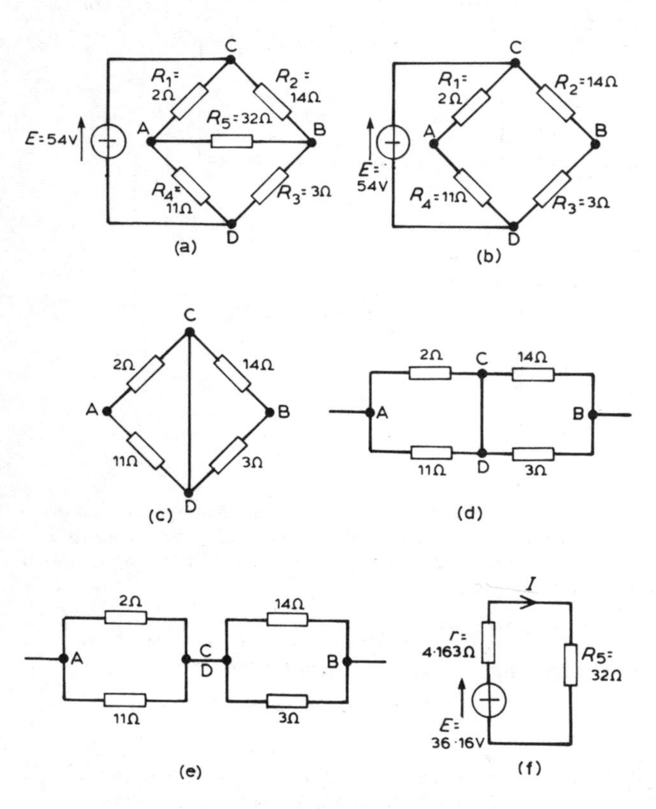

Fig 29

(iii) Replacing the source of emf with a short–cicuit (i.e. zero internal resistance) gives the circuit shown in *Fig 29(c)*. The circuit is redrawn and simplified as shown in *Fig 29(d) and (e)*, from which the resistance between terminals A and B,

$$r = \frac{2\times11}{2+11} + \frac{14\times3}{14+3} = \frac{22}{13} + \frac{42}{17} = 1.692 + 2.471 = 4.163\ \Omega.$$

(iv) The equivalent Thévenin's circuit is shown in *Fig 29(f)*, from which,

current $I = \dfrac{E}{r+R_5} = \dfrac{36.16}{4.163+32} = 1$ A.

Hence the current in the 32 Ω resistor of Fig 29(a) is 1 A, flowing from A to B

NORTON'S THEOREM

Problem 9 Use Norton's theorem to determine the current flowing in the 10 Ω resistance for the circuit shown in *Fig 30(a)*.

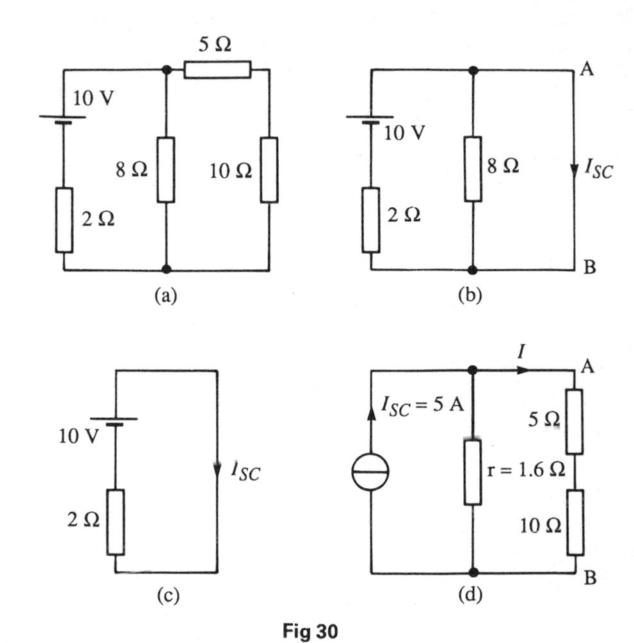

Fig 30

Following the procedure of para. 8:

(i) The branch containing the 10 Ω resistance is short–circuited as shown in *Fig 30(b)*.

(ii) *Fig 30(c)* is equivalent to *Fig 30(b)*. Hence $I_{sc} = \dfrac{10}{2} = 5$ A

(iii) If the 10 V source of emf is removed from *Fig 30(b)* the resistance 'looking-in' at a break made between A and B is given by:

$$r = \frac{2 \times 8}{2 + 8} = 1.6\ \Omega$$

(iv) From the Norton equivalent network shown in *Fig 30(d)* the current in the 10 Ω resistance, by current division, is given by:

$$I = \left(\frac{1.6}{1.6 + 5 + 10}\right)(5) = \mathbf{0.482\ A},$$

as obtained previously in *problem 3* using Thévenin's theorem.

Problem 10 Use Norton's theorem to determine the current I flowing in the 4 Ω resistance shown in *Fig 31(a)*.

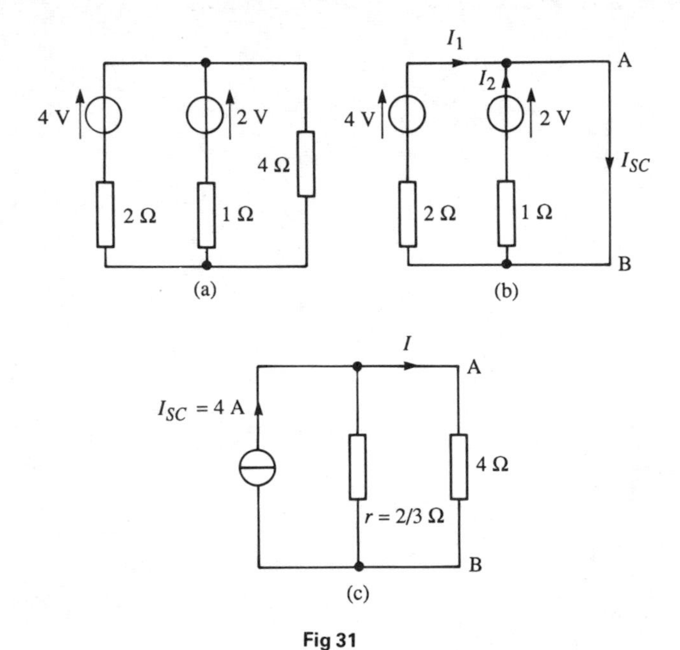

Fig 31

Following the procedure of para. 8:

(i) The 4 Ω branch is short–circuited as shown in *Fig 31(b)*.

(ii) From *Fig 31(b)*, $I_{sc} = I_1 + I_2 = \frac{4}{2} + \frac{2}{1} = 4$ A.

(iii) If the sources of emf are removed the resistance 'looking-in' at a break made between A and B is given by:

$$r = \frac{2 \times 1}{2 + 1} = \frac{2}{3}\ \Omega$$

(iv) From the Norton equivalent network shown in *Fig 31(c)* the current in the 4 Ω resistance is given by:

$I = \left(\frac{\frac{2}{3}}{\frac{2}{3}+4}\right)(4) = $ **0.571 A**, as obtained previously in *problems 1 and 5* using the theorems of superposition and Thévenin.

Problem 11 Determine the current in the 5 Ω resistance of the network shown in *Fig 32(a)* using Norton's theorem. Hence find the currents flowing in the other two branches.

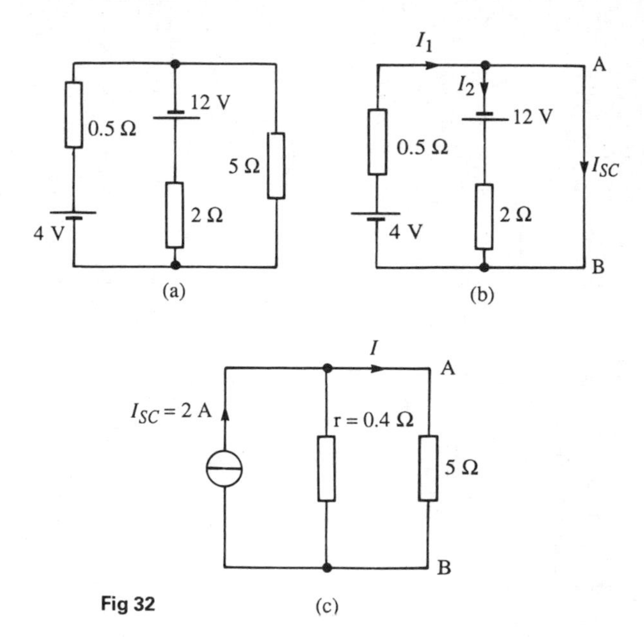

Fig 32

Following the procedure of para. 8:

(i) The 5 Ω branch is short–circuited as shown in *Fig 32(b)*.

(ii) From *Fig 32(b)*, $I_{sc} = I_1 - I_2 = \frac{4}{0.5} - \frac{12}{2} = 8 - 6 = 2$ A

(iii) If each source of emf is removed the resistance 'looking-in' at a break made between A and B is given by:

$$r = \frac{0.5 \times 2}{0.5 + 2} = 0.4\ \Omega$$

(iv) From the Norton equivalent network shown in *Fig 32(c)* the current in the 5 Ω resistance is given by:

$I = \left(\frac{0.4}{0.4+5}\right)(2) = $ **0.148 A**, as obtained previously in *problem 6* using Thévenin's theorem.

The currents flowing in the other two branches are obtained in the same way as in *problem 6*. Hence the current flowing from the 4 V source is **6.52 A** and the current flowing from the 12 V source is **6.37 A**

Problem 12 Use Norton's theorem to determine the current flowing in the 3 Ω resistance of the network shown in *Fig 33(a)*. The voltage source has negligible internal resistance.

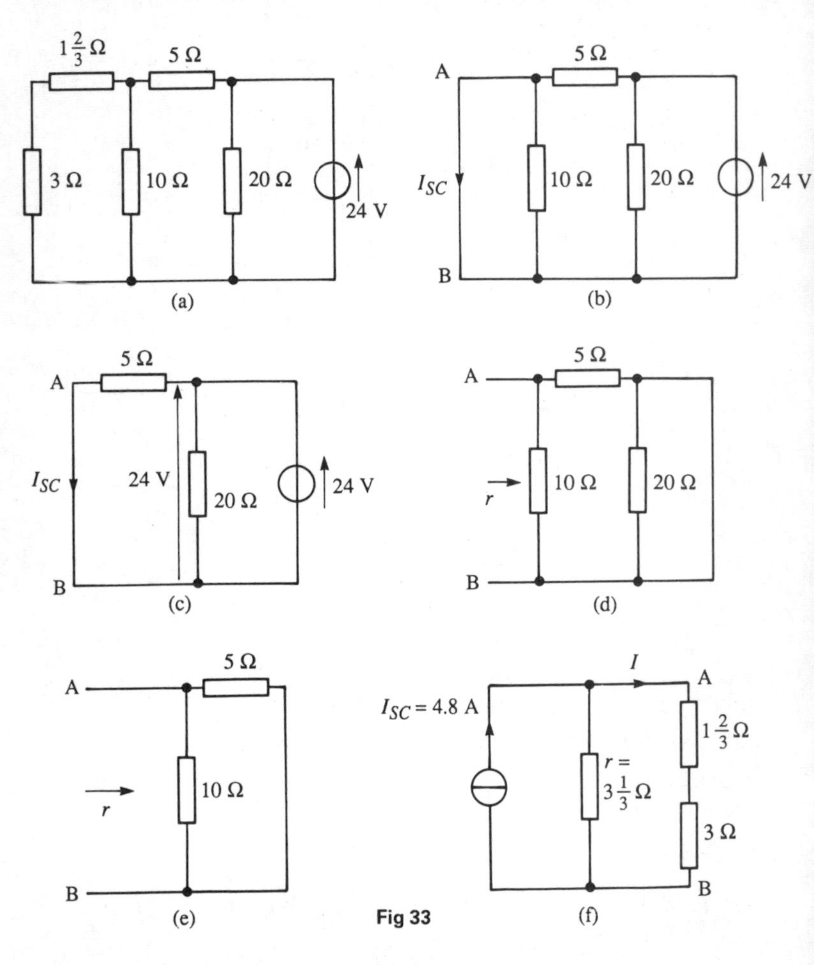

Fig 33

Following the procedure of para. 8:

(i) The branch containing the 3 Ω resistance is short–circuited as shown in *Fig 33(b)*.

(ii) From the equivalent circuit shown in *Fig 33(c)*, $I_{sc} = \frac{24}{5} = 4.8$ A.

(iii) If the 24 V source of emf is removed the resistance 'looking-in' at a break made between A and B is obtained from *Fig 33(d)* and its equivalent circuit shown in *Fig 33(e)* and is given by:

$$r = \frac{10 \times 5}{10 + 5} = \frac{50}{15} = 3\tfrac{1}{3}\ \Omega$$

(iv) From the Norton equivalent network shown in *Fig 33(f)* the current in the 3 Ω resistance is given by:

$I = \left(\frac{3\frac{1}{3}}{3\frac{1}{3} + 1\frac{2}{3} + 3}\right)(4.8) = \mathbf{2\ A}$, as obtained previously in *problem 7* using Thévenin's theorem.

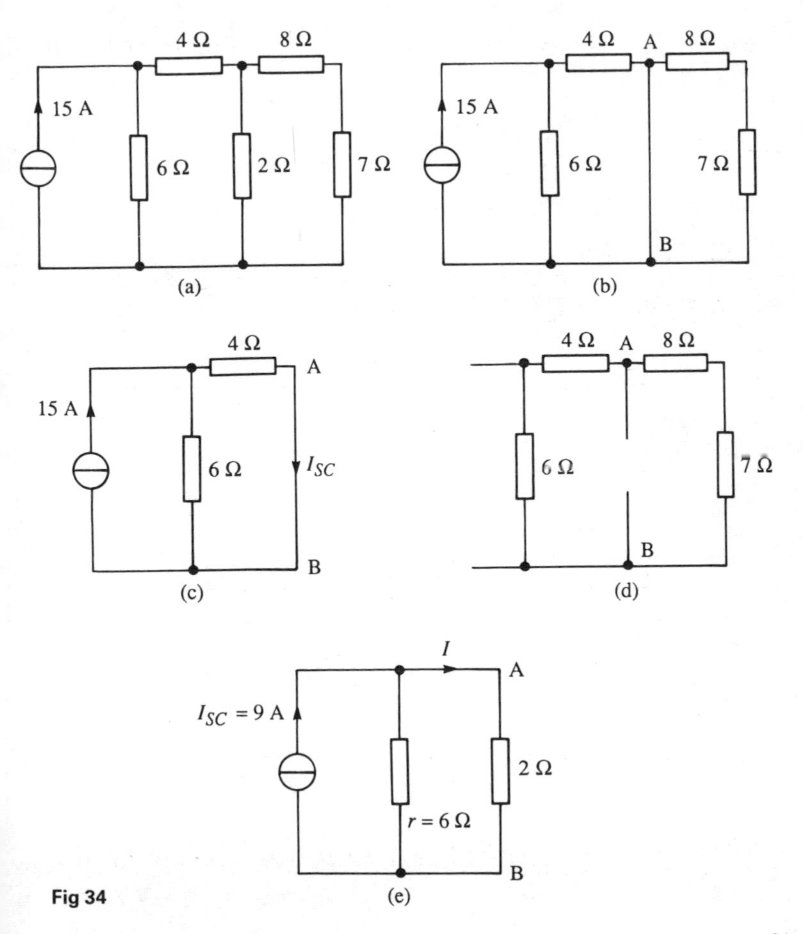

Fig 34

Problem 13 Determine the current flowing in the 2 Ω resistance in the network shown in *Fig 34(a)*.

Following the procedure of para. 8:

(i) The 2 Ω resistance branch is short–circuited as shown in *Fig 34(b)*.

(ii) *Fig 34(c)* is equivalent to *Fig 34(b)*. Hence $I_{sc} = \left(\frac{6}{6+4}\right)(15) = \mathbf{9\ A}$ by current division.

(iii) If the 15 A current source is replaced by an open–circuit then from *Fig 34(d)* the resistance 'looking-in' at a break made between A and B is given by (6 + 4) Ω in parallel with (8 + 7) Ω, i.e.

$$r = \frac{(10)(15)}{10 + 15} = \frac{150}{25} = 6\ \Omega$$

(iv) From the Norton equivalent network shown in *Fig 34(e)* the current in the 2 Ω resistance is given by:

$$I = \left(\frac{6}{6+2}\right)(9) = \mathbf{6.75\ A}$$

THÉVENIN AND NORTON EQUIVALENT NETWORKS

Problem 14 Convert the circuit shown in *Fig 35* to an equivalent Norton network.

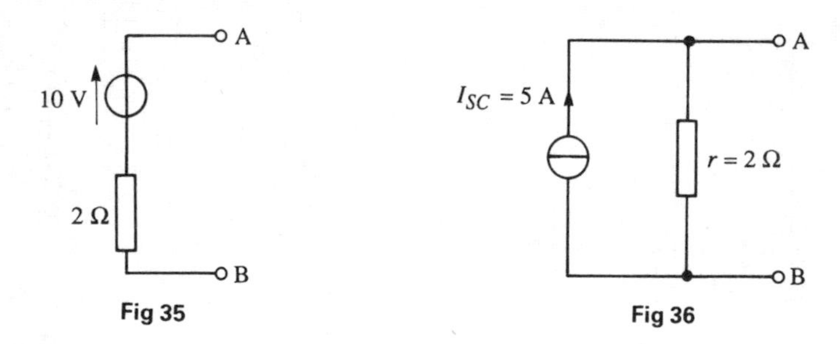

Fig 35

Fig 36

If terminals AB in *Fig 35* are short–circuited, the short–circuit current I_{sc} $= \frac{10}{2} = 5$ A.

The resistance 'looking-in' at terminals AB is 2 Ω. Hence the equivalent Norton network is as shown in *Fig 36*.

Problem 15 Convert the network shown in *Fig 37* to an equivalent Thévenin circuit.

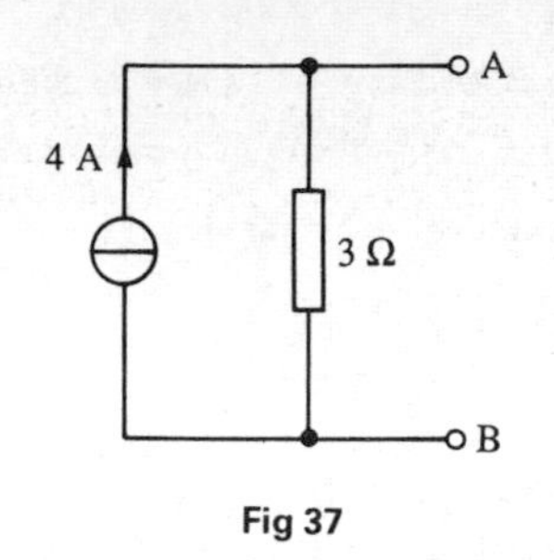

Fig 37

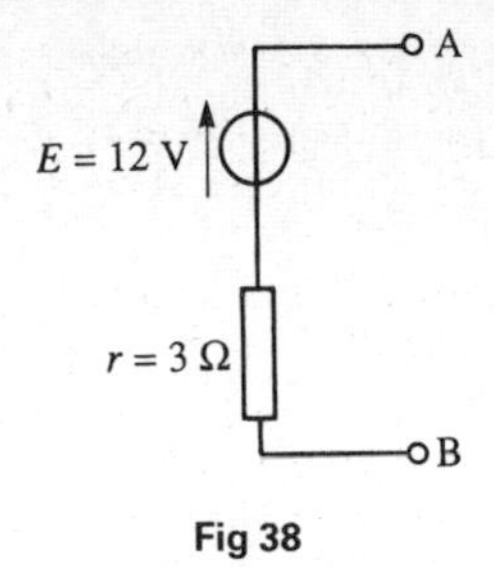

Fig 38

The open–circuit voltage E across terminals AB in *Fig 37* is given by:

$E = (I_{sc})(r) = (4)(3) = 12$ V.

The resistance 'looking-in' at terminals AB is 3 Ω. Hence the equivalent Thévenin circuit is as shown in *Fig 38*.

Problem 16 (a) Convert the circuit to the left of terminals AB in *Fig 39(a)* to an equivalent Thévenin circuit by initially converting to a Norton equivalent circuit. (b) Determine the current flowing in the 1.8 Ω resistor.

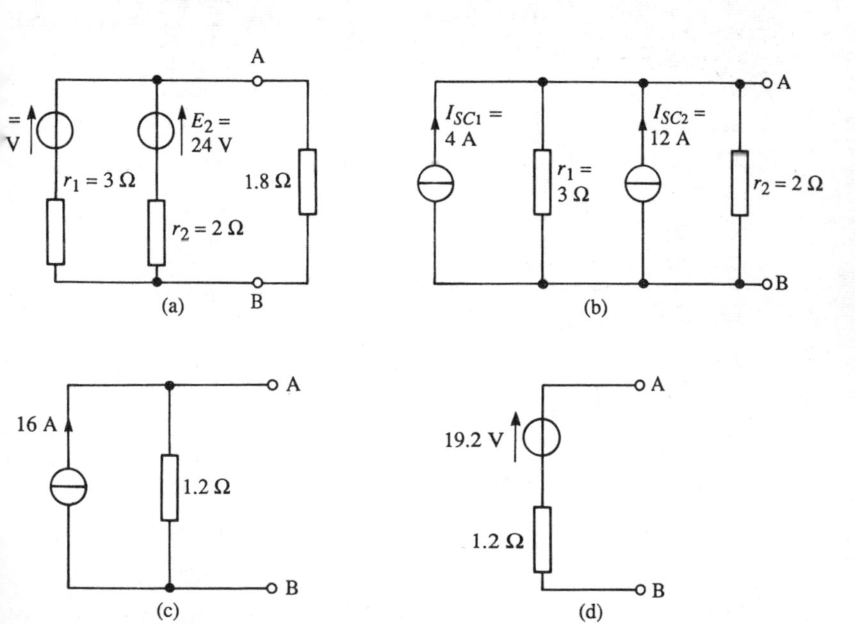

Fig 39

(a) For the branch containing the 12 V source, converting to a Norton equivalent circuit gives $I_{sc} = \frac{12}{3} = 4$ A and $r_1 = 3\ \Omega$. For the branch containing the 24 V source, converting to a Norton equivalent circuit gives $I_{sc_2} = \frac{24}{2} = 12$ A and $r_2 = 2\ \Omega$.
Thus *Fig 39(b)* shows a network equivalent to *Fig 39(a)*.
From *Fig 39(b)* the total short–circuit current is 4 + 12 = 16 A and the total resistance is given by: $\frac{3 \times 2}{3 + 2} = 1.2\ \Omega$.
Thus *Fig 39(b)* simplifies to *Fig 39(c)*.
The open–circuit voltage across AB of *Fig 39(c)*,

$$E = (16)(1.2) = 19.2 \text{ V}$$

and the resistance 'looking-in' at AB is 1.2 Ω. Hence the Thévenin equivalent circuit is as shown in *Fig 39(d)*.

(b) When the 1.8 Ω resistance is connected between terminals AB of *Fig 39(d)* the current I flowing is given by:

$$I = \frac{19.2}{1.2 + 1.8} = \mathbf{6.4\ A}$$

Problem 17 Determine by successive conversions between Thévenin and Norton equivalent networks a Thévenin equivalent circuit for terminals AB of *Fig 40(a)*. Hence determine the current flowing in the 200 Ω resistance.

For the branch containing the 10 V source, converting to a Norton equivalent network gives $I_{sc_1} = \frac{10}{2000} = 5$ mA and $r_1 = 2$ kΩ.
For the branch containing the 6 V source, converting to a Norton equivalent network gives $I_{sc_2} = \frac{6}{3000} = 2$ mA and $r_2 = 3$ kΩ.
Thus the network of *Fig 40(a)* converts to *Fig 40(b)*.
Combining the 5 mA and 2 mA current sources gives the equivalent network of *Fig 40(c)* where the short–circuit current for the original two branches considered is 7 mA and the resistance is $\frac{2 \times 3}{2 + 3} = 1.2$ kΩ.
Both of the Norton equivalent networks shown in *Fig 40(c)* may be converted to Thévenin equivalent circuits. The open–circuit voltage across CD is $(7 \times 10^{-3})(1.2 \times 10^{3}) = 8.4$ V and the resistance 'looking-in' at CD is 1.2 kΩ. The open–circuit voltage across EF is $(1 \times 10^{-3})(600) = 0.6$ V and the resistance 'looking-in' at EF is 0.6 kΩ. Thus *Fig 40(c)* converts to *Fig 40(d)*. Combining the two Thévenin circuits gives $E = 8.4 - 0.6 = 7.8$ V and the resistance $r = (1.2 + 0.6)$ kΩ $= 1.8$ kΩ.
Thus the Thévenin equivalent circuit for terminals AB of *Fig 40(a)* is as shown in *Fig 40(e)*.
Hence the current I flowing in a 200 Ω resistance connected between A and B is given by:

$$I = \frac{7.8}{1800 + 200} = \frac{7.8}{2000} = \mathbf{3.9\ mA}$$

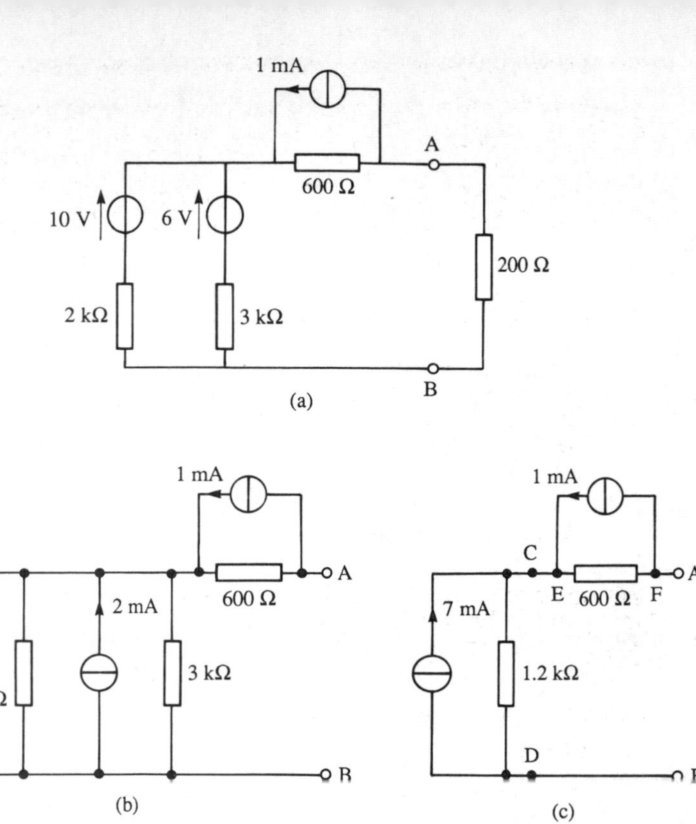
1 mA
A
600 Ω
10 V
6 V
200 Ω
2 kΩ
3 kΩ
B
(a)
1 mA
A
600 Ω
5 mA
2 mA
2 kΩ
3 kΩ
B
(b)
1 mA
C
A
E
600 Ω
F
7 mA
1.2 kΩ
D
B
(c)

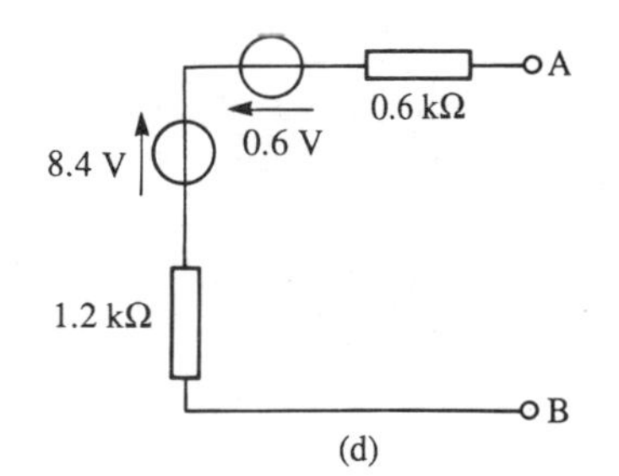
A
0.6 kΩ
0.6 V
8.4 V
1.2 kΩ
B
(d)

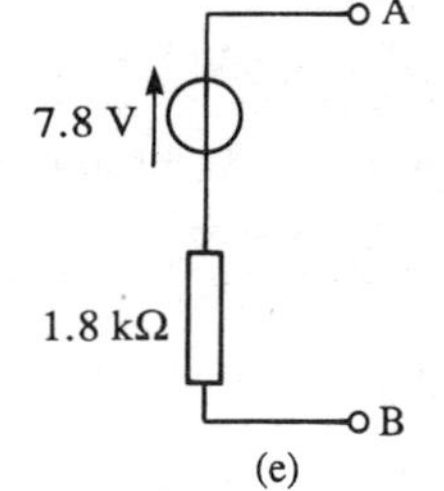
A
7.8 V
1.8 kΩ
B
(e)

Fig 40

Problem 18 The circuit diagram of *Fig 41* shows dry cells of source emf 6 V, and internal resistance 2.5 Ω. If the load resistance R_L is varied from 0 to 5 Ω in 0.5 Ω steps, calculate the power dissipated by the load in each case. Plot a graph of R_L (horizontally) against power (vertically) and determine the maximum power dissipated.

When $R_L = 0$, current $I = \frac{E}{r+R_L} = \frac{6}{2.5} = 2.4$ A and power dissipated in R_L, $P = I^2 R_L$, i.e. $P = (2.4)^2(0) = 0$ W.

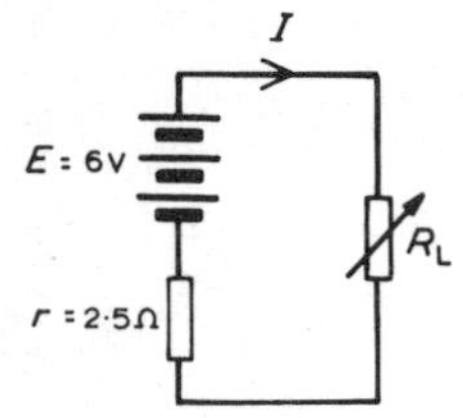

Fig 41

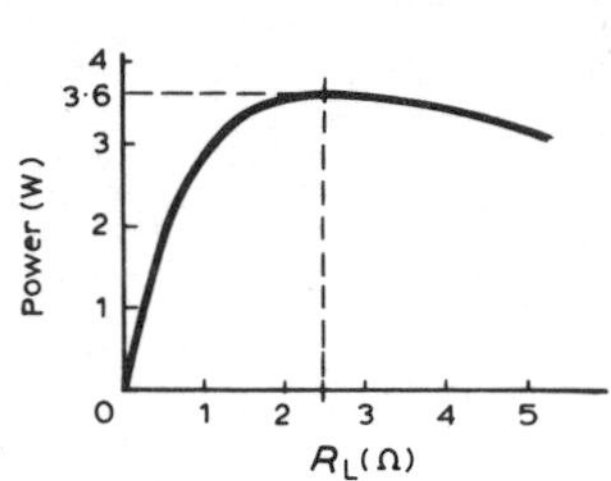

Fig 42

When $R_L = 0.5$ Ω, current $I = \frac{E}{r+R_L} = \frac{6}{2.5+0.5} = 2$ A

and $P = I^2 R_L = (2)^2(0.5) = 2$ W.

When $R_L = 1.0$ Ω, current $I = \frac{6}{2.5+1.0} = 1.714$ A and $P = (1.714)^2(1.0) = 2.94$ W.

With similar calculations the following table is produced:

R_L (Ω)	0	0.5	1.0	1.5	2.0	2.5	3.0	3.5	4.0	4.5	5.0
$I = \frac{E}{r+R_L}$	2.4	2.0	1.714	1.5	1.333	1.2	1.091	1.0	0.923	0.857	0.8
$P = I^2 R_L$ (W)	0	2.00	2.94	3.38	3.56	3.60	3.57	3.50	3.41	3.31	3.20

A graph of R_L against P is shown in *Fig 42*. **The maximum value of power is 3.60 W** which occurs when R_L is 2.5 Ω, i.e. maximum power occurs when $R_L = r$, which is what the maximum power transfer theorem states.

Problem 19 A d.c. source has an open–circuit voltage of 30 V and an internal resistance of 1.5 Ω. State the value of load resistance that gives maximum power dissipation and determine the value of this power.

The circuit diagram is shown in *Fig 43*. From the maximum power transfer theorem:

For maximum power dissipation, $R_L = r =$ **1.5 Ω**

From *Fig 43*, current $I = \frac{E}{r+R_L} = \frac{30}{1.5+1.5} = 10$ A

Power $P = I^2 R_L = (10)^2(1.5) =$ **150 W**

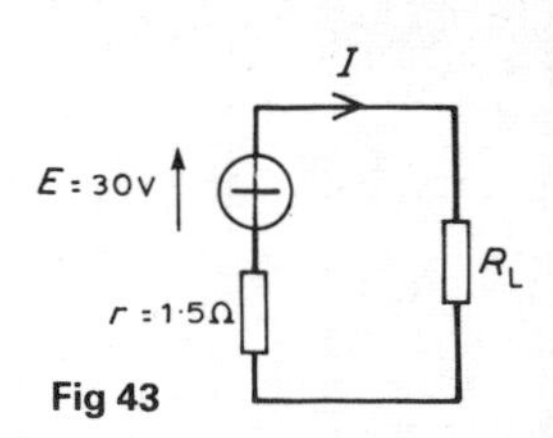

Fig 43

Problem 20 Find the value of the load resistor R_L shown in *Fig 44(a)* that gives maximum power dissipation and determine the value of this power.

Using the procedure for Thévenin's theorem given in para 5:

(i) Resistance R_L is removed from the circuit as shown in *Fig 44(b)*.

(ii) The p.d. across AB is the same as the p.d. across the 12 Ω resistor.

Hence $E = \left(\frac{12}{12+3}\right)(15) = 12$ V

(iii) Removing the source of emf gives the circuit of *Fig 44(c)*, from which,

resistance, $r = \frac{12 \times 3}{12+3} = \frac{36}{15} = 2.4\ \Omega$

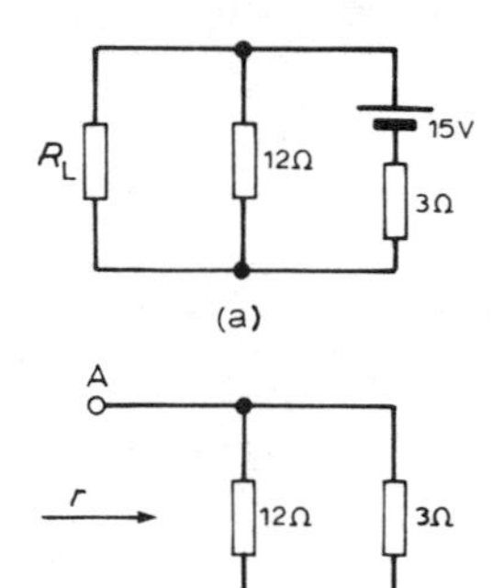

(a)

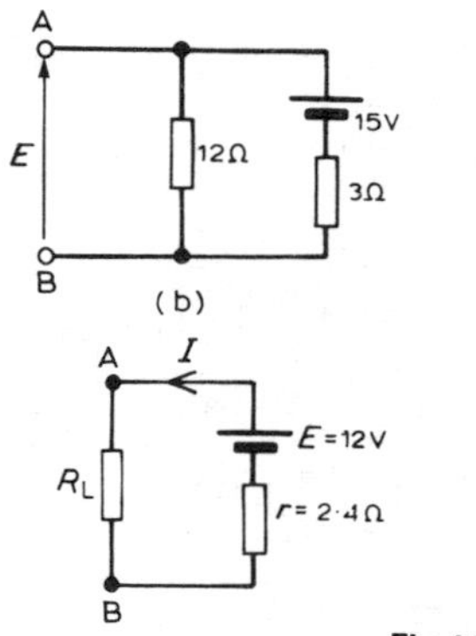

(b)

(c)

(d)

Fig 44

(iv) The equivalent Thévenin's circuit supplying terminals AB is shown in *Fig 44(d)*, from which

current, $I = \frac{E}{r+R_L}$

For maximum power, $R_L = r =$ **2.4 Ω**. Thus

current, $I = \frac{12}{2.4+2.4} = 2.5$ A

Power, P, dissipated in load R_L, $P = I^2 R_L = (2.5)^2(2.4) =$ **15 W**

Problem 21 A transformer having a turns ratio of 4:1 supplies a load of resistance 100 Ω. Determine the equivalent input resistance of the transformer.

From para. 12, the equivalent input resistance, $R_1 = \left(\frac{N_1}{N_2}\right)^2 R_L$

$= \left(\frac{4}{1}\right)^2 (100) =$ **1600 Ω**

Problem 22 The output stage of an amplifier has an output resistance of 112 Ω. Calculate the optimum turns ratio of a transformer which would match a load resistance of 7 Ω to the output resistance of the amplifier.

The circuit is shown in *Fig 45*.
The equivalent input resistance, R_1 of the transformer needs to be 112 Ω for maximum power transfer. From para. 12,

$$R_1 = \left(\frac{N_1}{N_2}\right)^2 R_L$$

Hence $\left(\frac{N_1}{N_2}\right)^2 = \frac{R_1}{R_L} = \frac{112}{7} = 16$

i.e. $\frac{N_1}{N_2} = \sqrt{(16)} = 4$

Hence the optimum turns ratio is 4:1

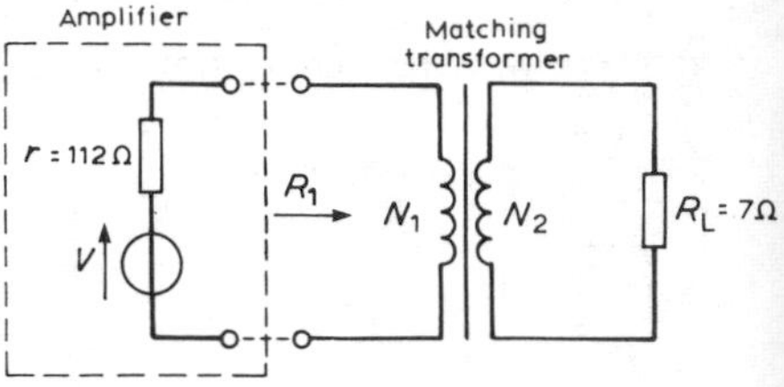

Fig 45

Problem 23 Determine the optimum value of load resistance for maximum power transfer if the load is connected to an amplifier of output resistance 150 Ω through a transformer with a turns ratio of 5:1.

The equivalent input resistance R_1 of the transformer needs to be 150 Ω for maximum power transfer.

From para. 12, $R_1 = \left(\frac{N_1}{N_2}\right)^2 R_L$, from which $R_L = R_1\left(\frac{N_2}{N_1}\right)^2 = 150\left(\frac{1}{5}\right)^2 = \mathbf{6\ \Omega}$

Problem 24 A single-phase, 220/1 760 V ideal transformer is supplied from a 220 V source through a cable of resistance 2 Ω. If the load across the secondary winding is 1.28 kΩ determine (a) the primary current flowing and (b) the power dissipated in the load resistor.

The circuit diagram is shown in *Fig 46*.

(a) Turns ratio $\frac{N_1}{N_2} = \frac{V_1}{V_2} = \frac{220}{1760} = \frac{1}{8}$.

Equivalent input resistance of the transformer,

$$R_1 = \left(\frac{N_1}{N_2}\right)^2 R_L$$

$$= \left(\frac{1}{8}\right)^2 (1.28 \times 10^3) = 20\ \Omega$$

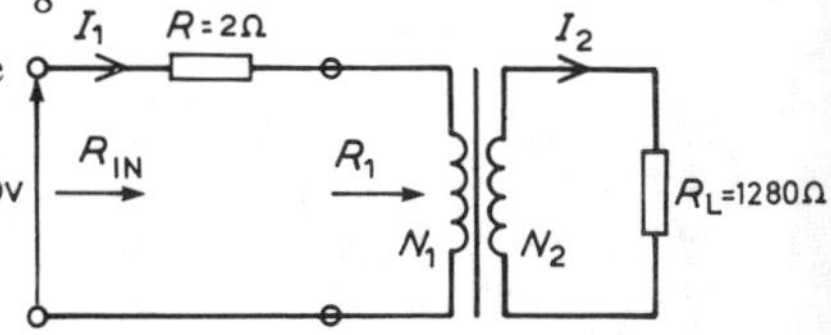

Fig 46

Total input resistance, $R_{IN} = R + R_1 = 2 + 20 = 22\ \Omega$

Primary current, $I_1 = \frac{V_1}{R_{IN}} = \frac{220}{22} = \mathbf{10\ A}$.

(b) For an ideal transformer $\frac{V_1}{V_2} = \frac{I_2}{I_1}$, from which $I_2 = I_1 \left(\frac{V_1}{V_2}\right) = 10\left(\frac{220}{1760}\right)$

$= 1.25$ A

Power dissipated in load resistor R_L, $P = I_2^2 R_L = (1.25)^2(1.28 \times 10^3)$

$=$ **2000 watts or 2 kW**

Problem 25 An a.c. source of 24 V and internal resistance 15 kΩ is matched to a load by a 25 : 1 ideal transformer. Determine (a) the value of the load resistance and (b) the power dissipated in the load.

The circuit diagram is shown in *Fig 47*

(a) For maximum power transfer R_1 needs to be equal to 15 kΩ.

From para. 12, $R_1 = \left(\frac{N_1}{N_2}\right)^2 R_L$, from which load resistance, $R_L = R_1\left(\frac{N_2}{N_1}\right)^2$

$= (15\,000)\left(\frac{1}{25}\right)^2 = 24\ \Omega$

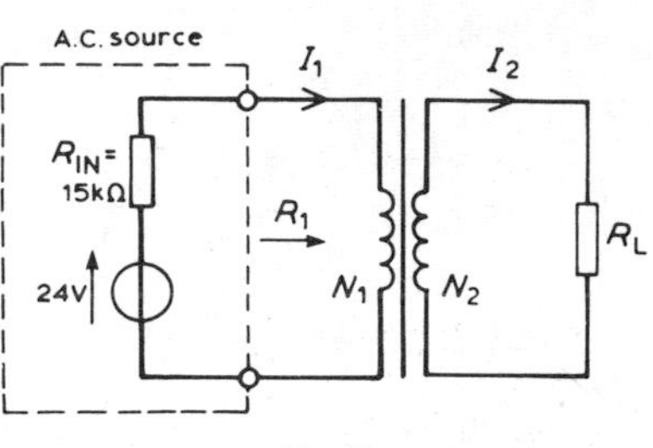

Fig 47

(b) The total input resistance when the source is connected to the matching transformer is $R_{IN} + R_1$, i.e. 15 kΩ+15kΩ = 30 kΩ.

Primary current, $I_1 = \frac{V}{30\,000} = \frac{24}{30\,000} = 0.8$ mA

$\frac{N_1}{N_2} = \frac{I_2}{I_1}$, from which $I_2 = I_1\left(\frac{N_1}{N_2}\right) = (0.8 \times 10^{-3})\left(\frac{25}{1}\right) = 20 \times 10^{-3}$ A

Power dissipated in the load R_L, $P = I_2^2 R_L = (20 \times 10^{-3})^2(24)$

$= 9600 \times 10^{-6}$ W $=$ **9.6 mW**

C. FURTHER PROBLEMS ON CIRCUIT THEOREMS

SHORT ANSWER PROBLEMS

1 Name two laws and three theorems which may be used to find currents and p.d.'s in electrical circuits.
2 State, in your own words, the superposition theorem.
3 State, in your own words, Thévenin's theorem.
4 State, in your own words, Norton's theorem.
5 State the maximum power transfer theorem.
6 What does 'resistance matching' mean?
7 State a practical source to which the maximum power transfer theorem is applicable.
8 Derive a formula for the equivalent resistance of a transformer having a turns ratio $N_1 : N_2$ and load resistance R_L.

MULTI-CHOICE PROBLEMS (answers on page 191)

1 For the circuit shown in *Fig 48*, the internal resistance, r, is given by:

(a) $\dfrac{I}{V-E}$; (b) $\dfrac{V-E}{I}$ (c) $\dfrac{I}{E-V}$; (d) $\dfrac{E-V}{I}$

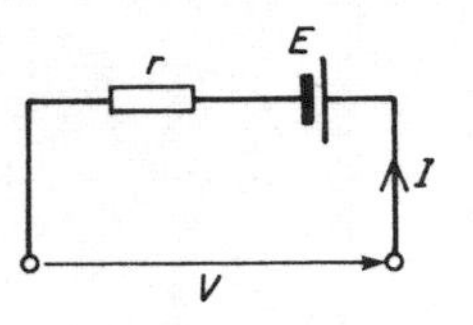

Fig 48

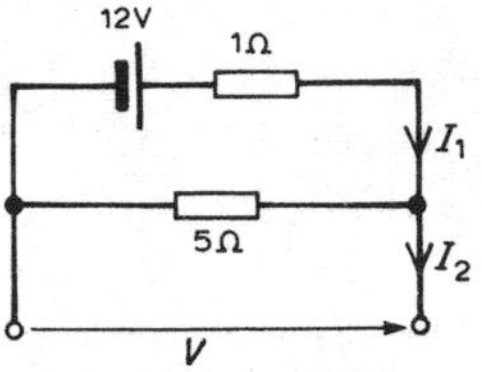

Fig 49

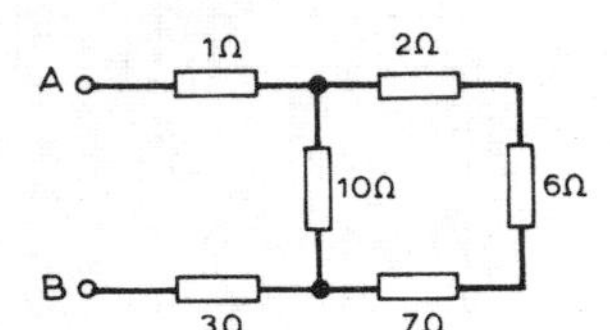

Fig 50

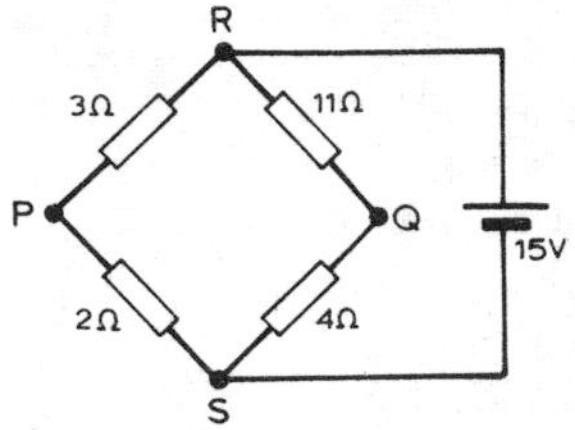

Fig 51

2 For the circuit shown in *Fig 49*, voltage V is:
(a) 12 V; (b) 2 V; (c) 10 V; (d) 0 V.

3 For the circuit shown in *Fig 49*, current I_1 is: (a) 2 A; (b) 14.4 A; (c) 0.5 A; (d) 0 A.

4 For the circuit shown in *Fig 49*, current I_2 is:
(a) 2 A; (b) 14.4 A; (c) 0.5 A; (d) 0 A.

5 The equivalent resistance across terminals AB of *Fig 50* is:
(a) 9.31 Ω; (b) 7.24 Ω;
(c) 10.0 Ω (d) 6.75 Ω.

6 With reference to *Fig 51*, which of the following statements is correct?
(a) $V_{PQ} = 2$ V; (b) $V_{PQ} = 15$ V
(c) When a load is connected between P and Q, current would flow from Q to P.
(d) $V_{PQ} = 20$ V.

7 In *Fig 51*, if the 15 V battery is replaced by a short-circuit, the equivalent resistance across terminal PQ is:
(a) 20 Ω; (b) 4.20 Ω; (c) 4.13 Ω; (d) 4.29 Ω.

8 For the circuit shown in *Fig 52*, maximum power transfer from the source is required. For this to be so, which of the following statements is true?
(a) $R_2 = 10$ Ω; (b) $R_2 = 30$ Ω; (c) $R_2 = 7.5$ Ω; (d) $R_2 = 15$ Ω.

9 The open-circuit voltage E across termins XY of *Fig 53* is:
(a) 0 V; (b) 20 V; (c) 4 V; (d) 16 V.

10 The maximum power transferred by the source in *Fig 54* is:
(a) 5 W; (b) 200 W; (c) 40 W; (d) 50 W.

11 A load is to be matched to an amplifier having an effective internal resistance of 10 Ω via a coupling transformer having a turns ratio of 1 : 10. The value of the load resistance for maximum power transfer is:
(a) 100 Ω; (b) 1 kΩ; (c) 100 mΩ; (d) 1 mΩ.

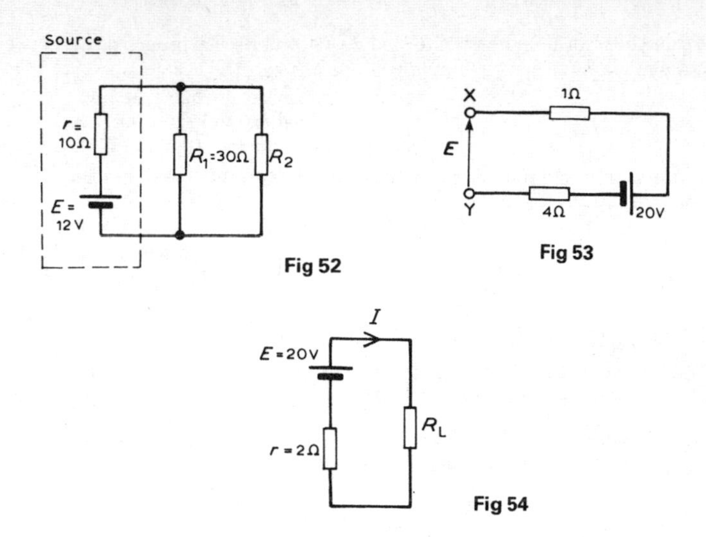

Fig 52

Fig 53

Fig 54

CONVENTIONAL PROBLEMS

Superposition theorem

1 Use the superposition theorem to find currents I_1, I_2 and I_3 of *Fig 55(a)*.
[$I_1 = 2$ A; $I_2 = 3$ A; $I_3 = 5$ A]

2 Use the superposition theorem to find the current in the 8 Ω resistor of *Fig 55(b)*.
[0.385 A]

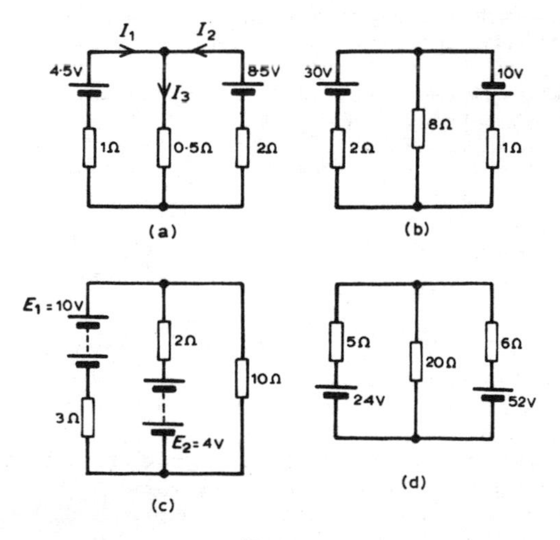

Fig 55

3 Use the superposition theorem to find the current in each branch of the network shown in *Fig 55(c)*.

[10 V battery discharges at 1.429 A;
4 V battery charges at 0.857 A;
Current through 10 Ω resistor is 0.572 A]

4 Use the superposition theorem to determine the current in each branch of the arrangement shown in *Fig 55(d)*.

[24 V battery charges at 1.664 A;
52 V battery discharges at 3.280 A;
Current in 20 Ω resistor is 1.616 A.]

Thévenin's theorem

5 Use Thévenin's theorem to find the current flowing in the 14 Ω resistor of the network shown in *Fig 56*. Find also the power dissipated in the 14 Ω resistor.
[0.434 A; 2.64 W]

6 Use Thévenin's theorem to find the current flowing in the 6 Ω resistor shown in *Fig 57* and the power dissipated in the 4 Ω resistor. [2.162 A; 42.07 W]

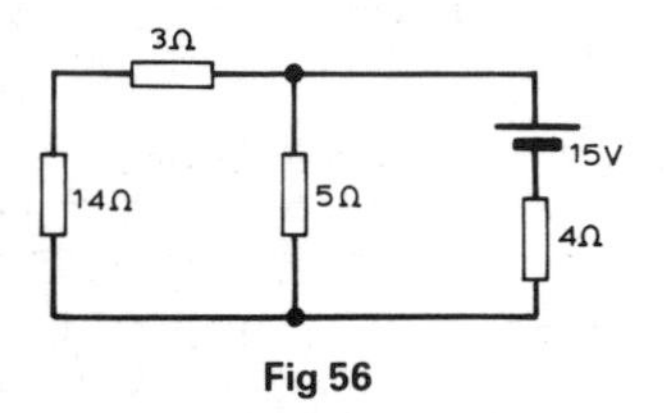

Fig 56

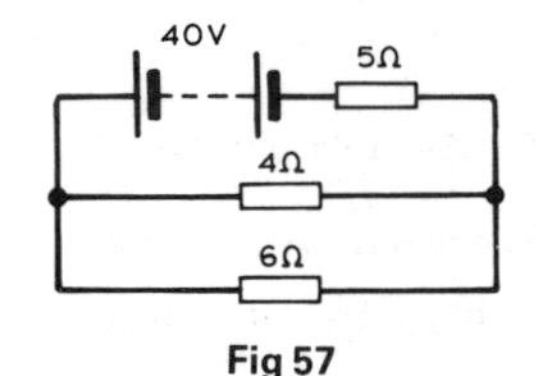

Fig 57

7 Repeat problems 1–4 using Thévenin's theorem.

8 In the network shown in *Fig 58* the battery has negligible internal resistance. Find, using Thévenin's theorem, the current flow in the 4 Ω resistor.
[0.918 A]

9 For the bridge network shown in *Fig 59*, find the current in the 5 Ω resistor, and its direction, by using Thévenin's theorem. [0.153 A from B to A]

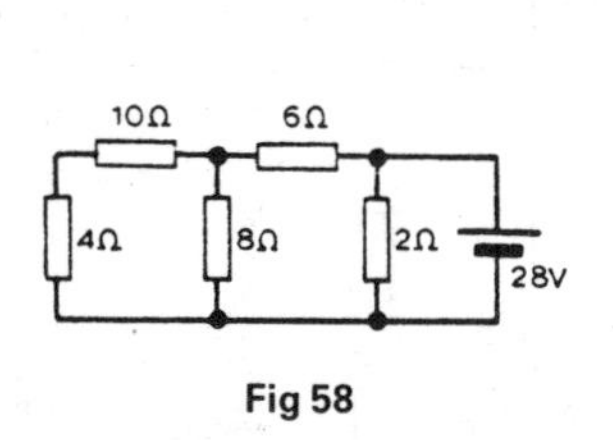

Fig 58

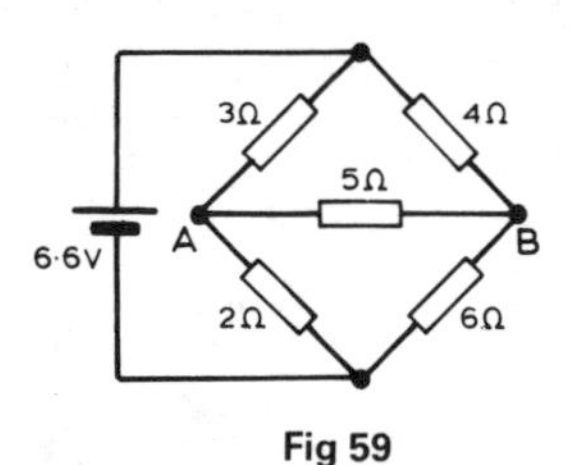

Fig 59

Norton's theorem

10 Repeat problems 1–6, 8 and 9 using Norton's theorem.

11 Determine the current flowing in the 6 Ω resistance of the network shown in *Fig 60* by using Norton's theorem. [2.5 mA]

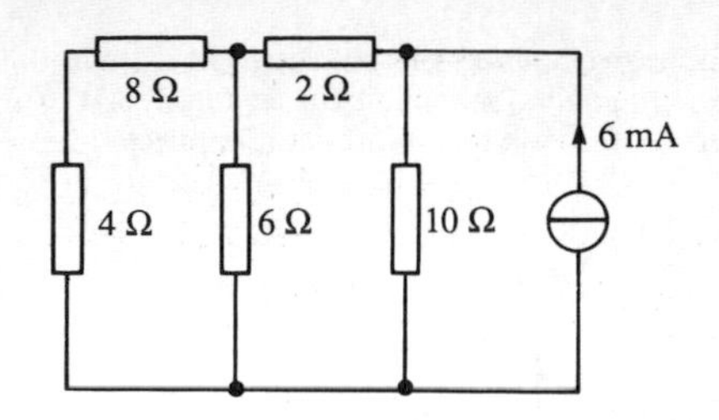

Fig 60

12 Convert the circuits shown in *Fig 61* to Norton equivalent networks.

[(a) $I_{sc} = 25$ A, $r = 2\ \Omega$
(b) $I_{sc} = 2$ mA, $r = 5\ \Omega$]

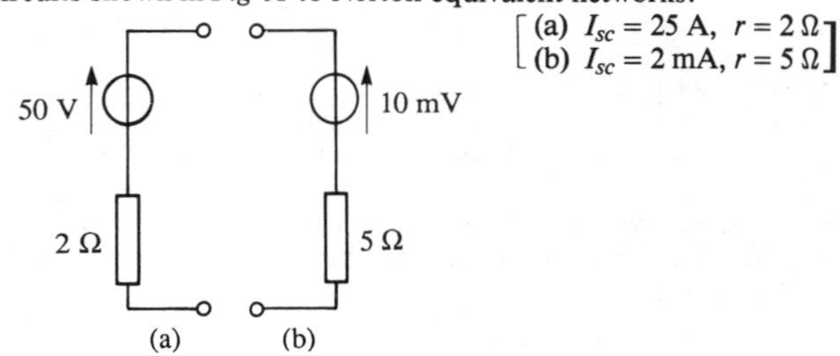

Fig 61

13 Convert the networks shown in *Fig 62* to Thévenin equivalent circuits.

[(a) $E = 20$ V, $r = 4\ \Omega$
(b) $E = 12$ mV, $r = 3\ \Omega$]

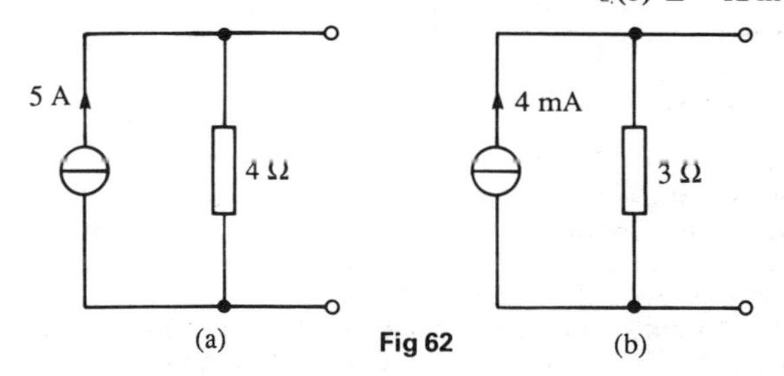

Fig 62

14 (a) Convert the network to the left of terminals AB in *Fig 63* to an equivalent Thévenin circuit by initially converting to a Norton equivalent network.
(b) Determine the current flowing in the 1.8 Ω resistance connected between A and B in *Fig 63*

[(a) $E = 18$ V, $r = 1.2\ \Omega$
(b) 6A]

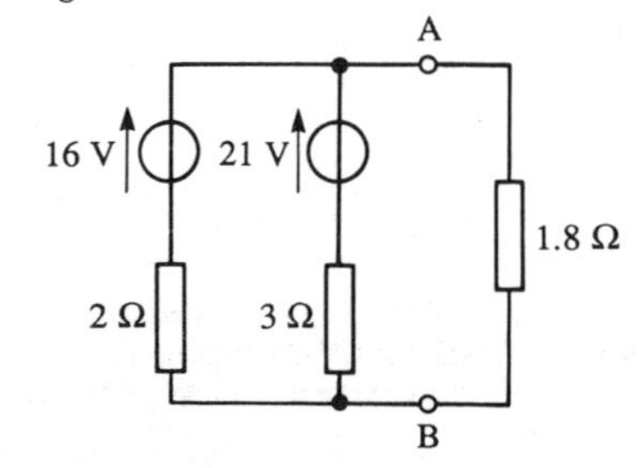

Fig 63

15 Determine, by successive conversions between Thévenin and Norton equivalent networks, a Thévenin equivalent circuit for terminals AB of *Fig 64*. Hence determine the current flowing in a 6 Ω resistor connected between A and B.
[$E = 9\frac{1}{3}$ V, $r = 1$ Ω; $1\frac{1}{3}$ V A]

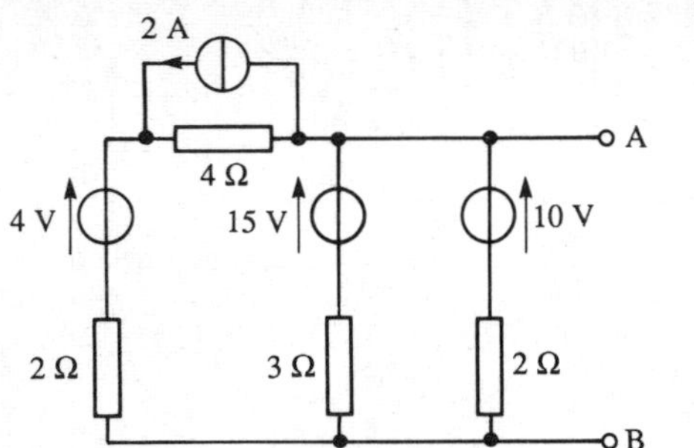

Fig 64

16 For the network shown in *Fig 65*, convert each branch containing a voltage source to its Norton equivalent and hence determine the current flowing in the 5 Ω resistance. [1.22 A]

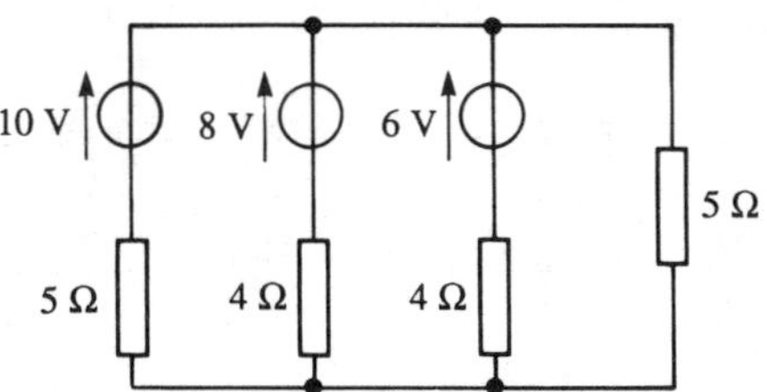

Fig 65

Maximum power transfer theorem and resistance matching

17 A d.c. source has an open-circuit voltage of 20 V and an internal resistance of 2 Ω. Determine the value of the load resistance that gives maximum power dissipation. Find the value of this power. [2 Ω; 50 W]

18 Determine the value of the load resistance R_L shown in *Fig 66* that gives maximum power dissipation and find the value of the power.
[$R_L = 1.6$ Ω; $P = 57.6$ W]

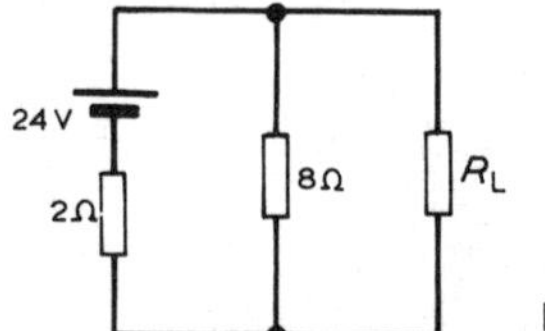

Fig 66

19 A transformer having a turns ratio of 8:1 supplies a load of resistance 50 Ω Determine the equivalent input resistance of the transformer. [3.2 kΩ]

20 What ratio of transformer is required to make a load of resistance 30 Ω appear to have a resistance of 270 Ω? [3:1]

21 State the maximum power transfer theorem.
Determine the optimum value of load resistance for maximum power transfer if the load is connected to an amplifier of output resistance 147 Ω through a transformer with a turns ratio of 7:2. [12 Ω]

22 A single-phase, 240/2 880 V ideal transformer is supplied from a 240 V source through a cable of resistance 3 Ω. If the load across the secondary winding is 720 Ω determine (a) the primary current flowing and (b) the power dissipated in the load resistance. [(a) 30 A; (b) 4.5 kW]

23 A load of resistance 768 Ω is to be matched to an amplifier which has an effective output resistance of 12 Ω. Determine the turns ratio of the coupling transformer. [1:8]

24 An a.c. source of 20 V and internal resistance 20 kΩ is matched to a load by a 16:1 single-phase transformer. Determine (a) the value of the load resistance and (b) the power dissipated in the load. [(a) 78.13 Ω; (b) 20 mW]

2 Single-phase series a.c. circuits

A. MAIN POINTS CONCERNED WITH SINGLE-PHASE SERIES A.C. CIRCUITS

1 In a purely resistive a.c. circuit, the current I_R and applied voltage V_R are in phase. See *Fig 1(a)*.

2 In a purely inductive a.c. circuit, the current I_L **lags** the applied voltage V_L by 90° (i.e. $\pi/2$ rads). See *Fig 1(b)*.

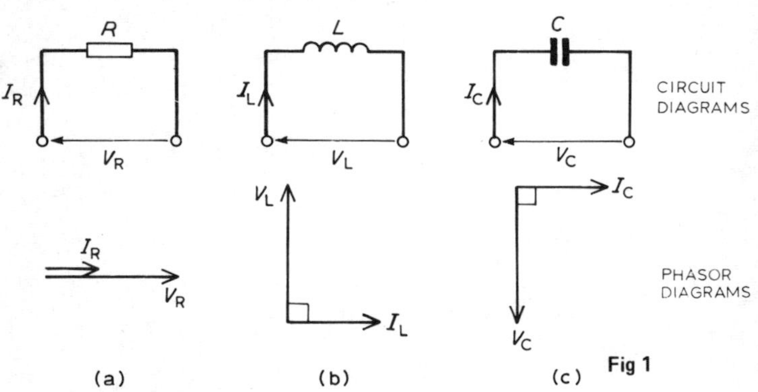

Fig 1

3 In a purely capacitive a.c. circuit, the current I_C **leads** the applied voltage V_C by 90° (i.e. $\pi/2$ rads). See *Fig 1(c)*.

4 In a purely inductive circuit the opposition to the flow of alternating current is called the **inductive reactance,** X_L.

$$X_L = \frac{V_L}{I_L} = 2\pi f L \ \Omega$$

where f is the supply frequency, in hertz, and L is the inductance, in henry's.

5 In a purely capacitive circuit the opposition to the flow of alternating current is called the **capacitance reactance,** X_C.

$$X_C = \frac{V_C}{I_C} = \frac{1}{2\pi f C} \ \Omega$$

where C is the capacitance in farads.

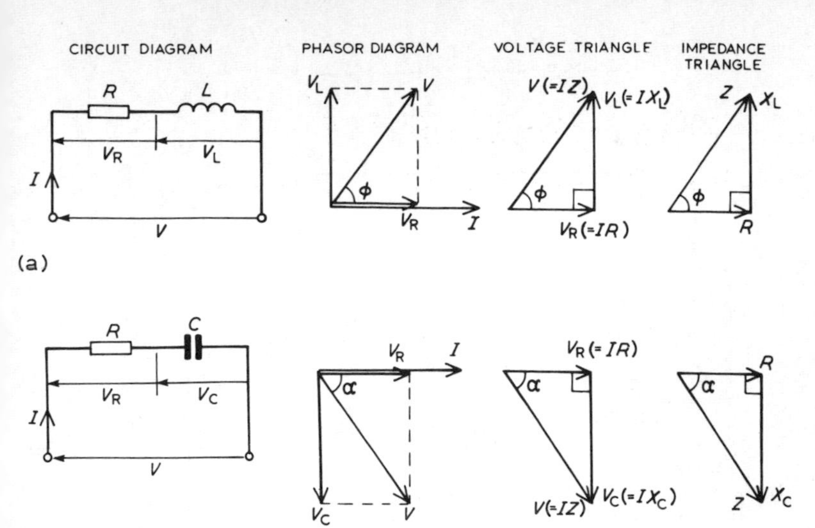

(b) **Fig 2**

6 In an a.c. circuit containing inductance L and resistance R, the applied voltage V is the phasor sum of V_R and V_L (see *Fig 2(a)*) and thus the current I lags the applied voltage V by an angle lying between 0° and 90° (depending on the values of V_R and V_L), shown as angle ϕ. In any a.c. series circuit the current is common to each component and is thus taken as the reference phasor.

7 In an a.c. series circuit containing capacitance C and resistance R, the applied voltage V is the phasor sum of V_R and V_C (see *Fig 2(b)*) and thus the current I leads the applied voltage V by an angle lying between 0° and 90° (depending on the values of V_R and V_C), shown as angle α.

8 In an a.c. circuit, the ratio applied voltage (V)/current (I) is called the **impedance** Z, i.e. $Z = \frac{V}{I}\,\Omega$

9 From the phasor diagrams of *Fig 2*, the **'voltage triangles'** are derived.

(a) For the R–L circuit: $V = \sqrt{(V_R^2 + V_L^2)}$ (by Pythagoras' theorem)

and $\tan\phi = \frac{V_L}{V_R}$ (by trigonometric ratios)

(b) For the R–C circuit: $V = \sqrt{(V_R^2 + V_C^2)}$

and $\tan\alpha = \frac{V_C}{V_R}$

10 If each side of the voltage triangles in *Fig 2* is divided by current I then the **'impedance triangles'** are derived.

(a) For the R–L circuit: $Z = \sqrt{(R^2 + X_L^2)}$

$\tan\phi = \frac{X_L}{R}$, $\sin\phi = \frac{X_L}{Z}$ and $\cos\phi = \frac{R}{Z}$

(b) For the R–C circuit: $Z = \sqrt{(R^2 + X_C^2)}$

$\tan\alpha = \frac{X_C}{R}$, $\sin\alpha = \frac{X_C}{Z}$ and $\cos\alpha = \frac{R}{Z}$

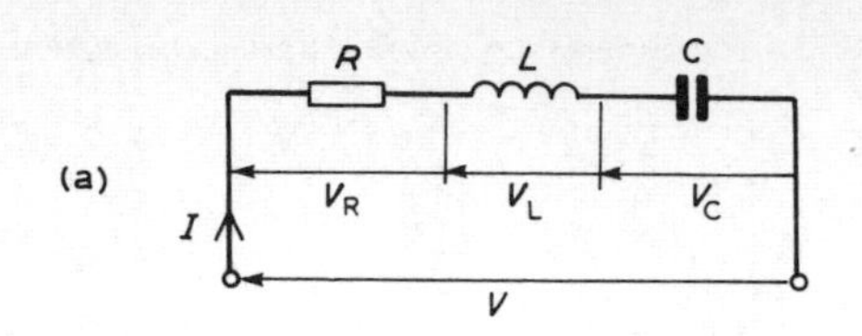

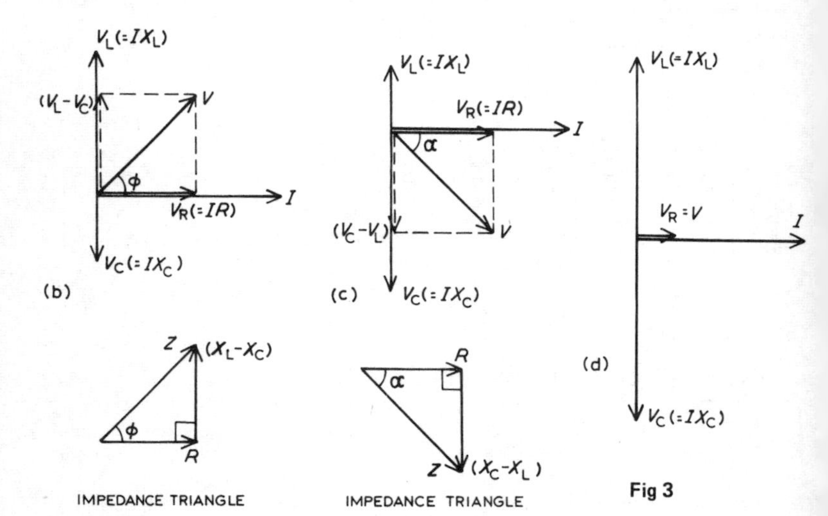

Fig 3

11 In an a.c. series circuit containing resistance R, inductance L and capacitance C, the applied voltage V is the phasor sum of V_R, V_L and V_C (see *Fig 3*). V_L and V_C are anti-phase, i.e. displaced by 180°, and there are three phasor diagrams possible—each depending on the relative values of V_L and V_C.

12 When $X_L > X_C$ (*Fig 3(b)*): $Z = \sqrt{[R^2+(X_L-X_C)^2]}$

$$\text{and } \tan\phi = \frac{(X_L-X_C)}{R}$$

13 When $X_C > X_L$ (*Fig 3(c)*): $Z = \sqrt{[R^2+(X_C-X_L)^2]}$

$$\text{and } \tan\alpha = \frac{(X_C-X_L)}{R}$$

14 When $X_L = X_C$ (*Fig 3(d)*), the applied voltage V and the current I are in phase. This effect is called **series resonance**. At resonance:

(i) $V_L = V_C$

(ii) $Z = R$ (i.e. the minimum circuit impedance possible in an L-C-R circuit).

(iii) $I = \frac{V}{R}$ (i.e. the maximum current possible in an L-C-R circuit)

(iv) Since $X_L = X_C$, then $2\pi f_r L = \frac{1}{2\pi f_r C}$, from which, $f_r = \frac{1}{2\pi\sqrt{(LC)}}$ Hz, where f_r is the resonant frequency.

(v) The series resonant circuit is often described as an **acceptor circuit** since it has its minimum impedance, and thus maximum current, at the resonant frequency.

(vi) Typical graphs of current I and impedance Z against frequency are shown in *Fig 4*.

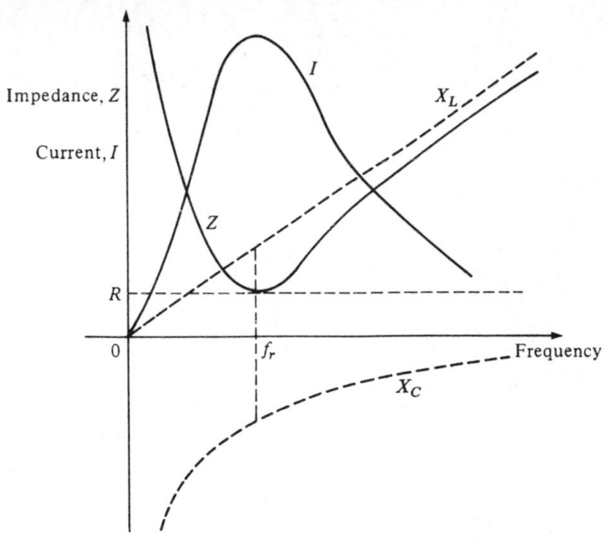

Fig 4 $|Z|$ and I plotted against frequency

15 At resonance, if R is small compared with X_L and X_C, it is possible for V_L and V_C to have voltages many times greater than the supply voltage (see *Fig 3(d)*).

$$\text{Voltage magnification at resonance} = \frac{\text{voltage across } L \text{ (or } C)}{\text{supply voltage } V}$$

This ratio is a measure of the quality of a circuit (as a resonator or tuning device) and is called the **Q-factor**.

$$\text{Hence Q-factor} = \frac{V_L}{V} = \frac{IX_L}{IR} = \frac{X_L}{R} = \frac{2\pi f_r L}{R}$$

$$\left(\text{Alternatively, Q-factor} = \frac{V_C}{V} = \frac{IX_C}{IR} = \frac{X_C}{R} = \frac{1}{2\pi f_r CR}\right)$$

$$\text{At resonance } f_r = \frac{1}{2\pi\sqrt{(LC)}} \quad \text{i.e. } 2\pi f_r = \frac{1}{\sqrt{(LC)}}$$

$$\text{Hence Q-factor} = \frac{2\pi f_r L}{R} = \frac{1}{\sqrt{(LC)}}\left(\frac{L}{R}\right) = \frac{1}{R}\sqrt{\left(\frac{L}{C}\right)}$$

16 Figure 5 shows how current I varies with frequency in an R–L–C series circuit. At the resonant frequency f_r, current is a maximum value, shown as I_r. Also shown are the points A and B where the current is 0.707 of the maximum value at frequencies f_1 and f_2. The power delivered to the circuit is I^2R. At $I = 0.707\ I_r$, the power is $(0.707\ I_r)^2\ R = 0.5\ I_r^2R$, i.e., half the power that occurs at frequency f_r. The points corresponding to f_1 and f_2 are called the **half-power points**. The distance between these points, i.e. $(f_2 - f_1)$, is called the **bandwidth**.

It may be shown that $Q = \frac{f_r}{f_2 - f_1}$

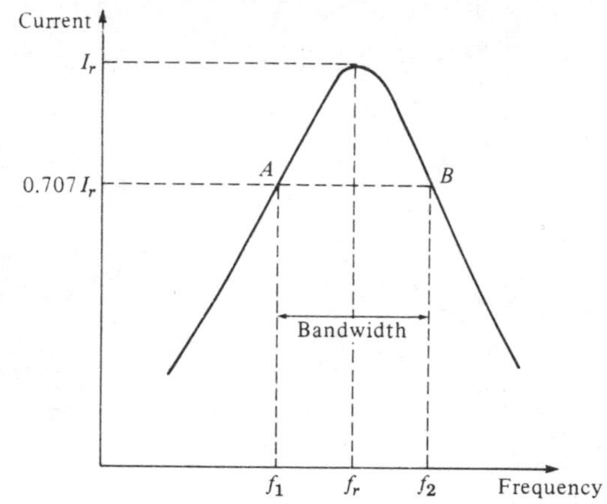

Fig 5

Bandwidth and half-power points f_1, f_2

17 **Selectivity** is the ability of a circuit to respond more readily to signals of a particular frequency to which it is tuned than to signals of other frequencies. The response becomes progressively weaker as the frequency departs from the resonant frequency. Discrimination against other signals becomes more pronounced as circuit losses are reduced, i.e., as the *Q*-factor is increased. Thus $Q_r = f_r/(f_2 - f_1)$ is a measure of the circuit selectivity in terms of the points on each side of resonance where the circuit current has fallen to 0.707 of its maximum value reached at resonance. The higher the *Q*-factor, the narrower the bandwidth and the more selective is the circuit. Circuits having high *Q*-factors (say, in the order of 100 to 300) are therefore useful in communications engineering. A high *Q*-factor in a series power circuit has disadvantages in that it can lead to dangerously high voltages across the insulation and may result in electrical breakdown.

For example, suppose that the working voltage of a capacitor is stated as 1 kV and is used in a circuit having a supply voltage of 240 V. The maximum value of the supply will be $\sqrt{2}(240)$, i.e., 340 V. The working voltage of the capacitor would appear to be ample. However, if the *Q*-factor is, say, 10, the voltage across the capacitor will reach 2.4 kV. Since the capacitor is rated only at 1 kV, dielectric breakdown is more than likely to occur.

Low *Q*-factors, say, in the order of 5 to 25, may be found in power transformers using laminated iron cores.

A capacitor-start induction motor, as used in domestic appliances such as washing-machines and vacuum-cleaners, having a *Q*-factor as low as 1.5 at starting would result in a voltage across the capacitor 1.5 times that of the supply voltage; hence the cable joining the capacitor to the motor would require extra insulation.

18 For series-connected impedances the total circuit impedance can be represented as a single *L-C-R* circuit by combining all values of resistance together, all values of inductance together and all values of capacitance together (remembering that for series connected capacitors $\frac{1}{C} = \frac{1}{C_1} + \frac{1}{C_2} + \ldots\ldots$). For example, the circuit of *Fig 6(a)* showing three impedances has an equivalent circuit of *Fig 6(b)*.

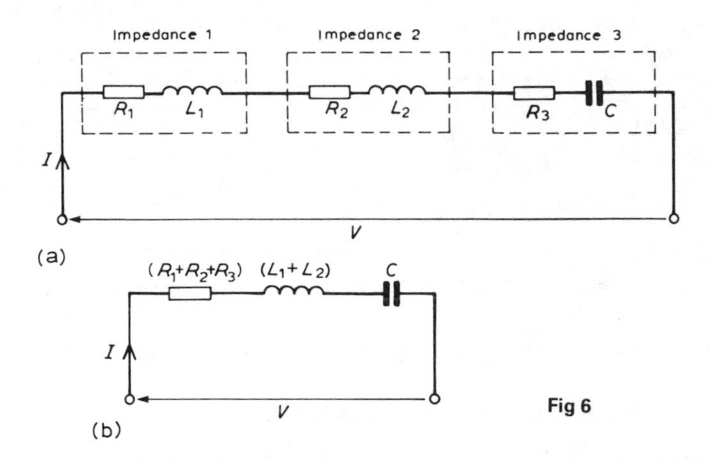

Fig 6

19 (a) For a purely resistive a.c. circuit, the average power dissipated, P, is given by $P = VI = I^2R = V^2/R$ watts (V and I being r.m.s. values).

(b) For a purely inductive or a purely capacitive a.c. circuit, the average power is zero.

(c) For an *R-L*, *R-C* or *L-C-R* series a.c. circuit, the average power dissipated, P, is given by: $P = VI \cos \phi$ watts or $P = I^2R$ watts (V and I being r.m.s. values).

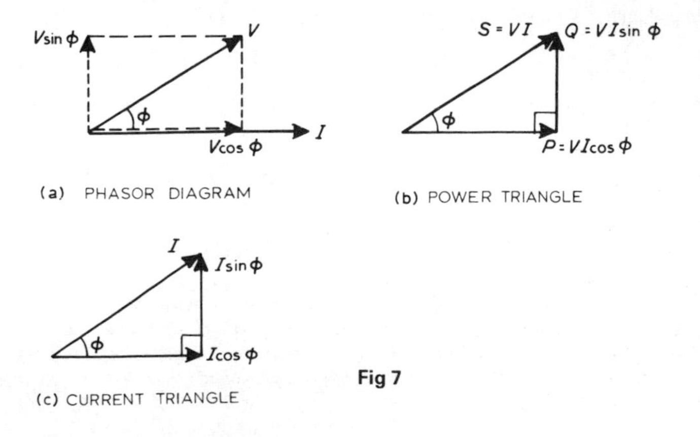

Fig 7

20 *Fig 7(a)* shows a phasor diagram in which the current I lags the applied voltage V by angle ϕ. The horizontal component of V is $V \cos \phi$ and the vertical component of V is $V \sin \phi$. If each of the voltage phasors is multiplied by I, *Fig 7(b)* is obtained and is known as the **'power triangle'**.

Apparent power, $S = VI$ voltamperes (VA)
True or active power, $P = VI \cos \phi$ watts (W)
Reactive power, $Q = VI \sin \phi$ reactive voltamperes (var)

21 If each of the phasors of the power triangle of *Fig 7(b)* is divided by V, *Fig 7(c)* is obtained and is known as the **'current triangle'**. The horizontal component of current, $I \cos \phi$, is called the **active** or the **in-phase component.** The vertical component of current, $I \sin \phi$, is called the **reactive** or the **quadrature component** (see *Problem 5*).

22 Power factor $= \dfrac{\text{True power } P}{\text{Apparent power } S}$

For sinusoidal voltages and currents, power factor $= \dfrac{P}{S} = \dfrac{VI \cos \phi}{VI}$

i.e. **p.f. = cos ϕ** $= \dfrac{R}{Z}$ (from *Fig 2*)

(The relationships stated in paras. 20 to 22 are also true when current I leads voltage V.)

B. WORKED PROBLEMS ON SERIES A.C. CIRCUITS

R-L A.C. circuits

Problem 1 A coil has a resistance of 15 Ω and an inductance 25.46 mH. Calculate (a) the inductive reactance, (b) the impedance and (c) the current taken from a 200 V, 50 Hz supply. Find also the phase angle between the supply voltage and current.

$R = 15\ \Omega$; $L = 25.46$ mH $= 25.46 \times 10^{-3}$ H; $f = 50$ Hz; $V = 200$ V

(a) Inductive reactance $X_L = 2\pi fL = 2\pi(50)(25.46 \times 10^{-3}) = 8.0\ \Omega$

(b) Impedance $Z = \sqrt{(R^2 + X_L^2)} = \sqrt{(15^2 + 8.0^2)} = \mathbf{17.0\ \Omega}$

(c) Current $I = \dfrac{V}{Z} = \dfrac{200}{17} = \mathbf{11.76\ A}$

The circuit and phasor diagrams and the voltage and impedance triangles are as shown in *Fig 2(a)*.

Since $\tan \phi = \dfrac{X_L}{R}$, $\phi = \arctan \dfrac{X_L}{R} = \arctan \dfrac{8.0}{15} = \mathbf{28^\circ\ 4'}$ **lagging**

('Lagging' infers that the current is 'behind' the voltage and is in the position shown on the phasor diagram, since phasors revolve anti-clockwise.)

Problem 2 A pure inductance of 1.273 mH is connected in series with a pure resistance of 30 Ω. If the frequency of the sinusoidal supply is 5 kHz and the p.d. across the 30 Ω resistor is 6 V, determine the value of the supply voltage and the voltage across the 1.273 mH inductance. Draw the phasor diagram.

The circuit is shown in *Fig 8(a)*.
Supply voltage, $V = IZ$

Current $I = \frac{V_R}{R} = \frac{6}{30} = 0.20$ A

Inductive reactance $X_L = 2\pi fL = 2\pi(5 \times 10^3)(1.273 \times 10^{-3}) = 40\ \Omega$
Impedance, $Z = \sqrt{(R^2 + X_L^2)} = \sqrt{(30^2 + 40^2)} = 50\ \Omega$
Supply voltage $V = IZ = (0.20)(50) =$ **10 V**
Voltage across the 1.273 mH inductance, $V_L = IX_L = (0.2)(40) =$ **8 V**

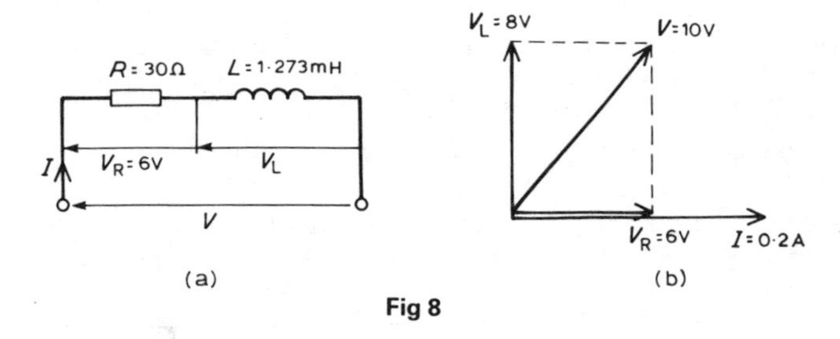

Fig 8

The phasor diagram is shown in *Fig 8(b)*. (Note that in a.c. circuits, the supply voltage is **not** the arithmetic sum of the p.d.'s across components but the **phasor** sum.)

Problem 3 A coil of inductance 159.2 mH and resistance 20 Ω is connected in series with a 60 Ω resistor to a 240 V, 50 Hz supply. Determine (a) the impedance of the circuit, (b) the current in the circuit, (c) the circuit phase angle, (d) the p.d. across the 60 Ω resistor and (e) the p.d. across the coil. (f) Draw the circuit phasor diagram showing all voltages.

The circuit diagram is shown in *Fig 9(a)*. When impedances are connected in series the individual resistances may be added to give the total circuit resistance. The equivalent circuit is thus shown in *Fig 9(b)*.

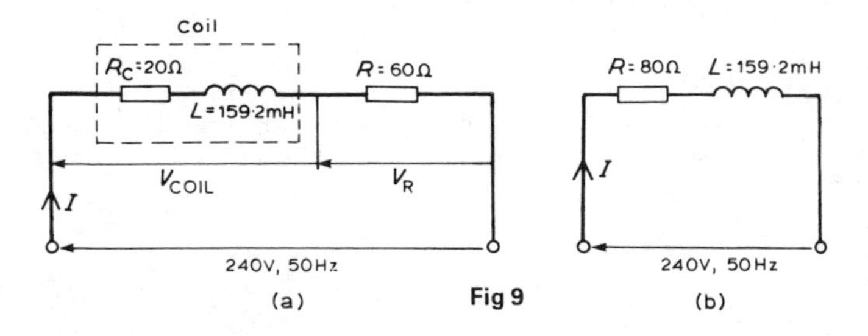

Fig 9

Inductive reactance $X_L = 2\pi fL = 2\pi(50)(159.2 \times 10^{-3}) = 50\ \Omega$
(a) Circuit impedance, $Z = \sqrt{(R_T^2 + X_L^2)} = \sqrt{(80^2 + 50^2)} = 94.34\ \Omega$

(b) Circuit current, $I = \frac{V}{Z} = \frac{240}{94.34}$ = **2.544 A**

(c) Circuit phase angle $\phi = \arctan \frac{X_L}{R}$ = arctan 50/80 = **32° lagging.**

From *Fig 9(a)*.

(d) $V_R = IR = (2.544)(60)$ = **152.6 V**

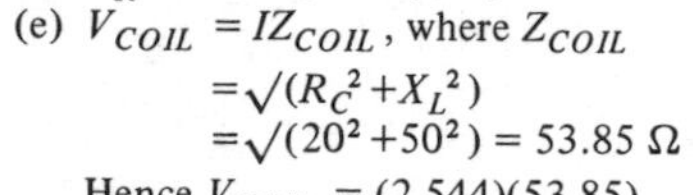

(e) $V_{COIL} = IZ_{COIL}$, where Z_{COIL}

$= \sqrt{(R_C^2 + X_L^2)}$

$= \sqrt{(20^2 + 50^2)} = 53.85\ \Omega$

Hence $V_{COIL} = (2.544)(53.85)$

= **137.0 V**

(f) For the phasor diagram, shown in *Fig 10*.

$V_L = IX_L = (2.544)(50) = 127.2$ V

$V_{R\,COIL} = IR_C = (2.544)(20) = 50.88$ V

The 240 V supply voltage is the phasor sum of V_{COIL} and V_R.

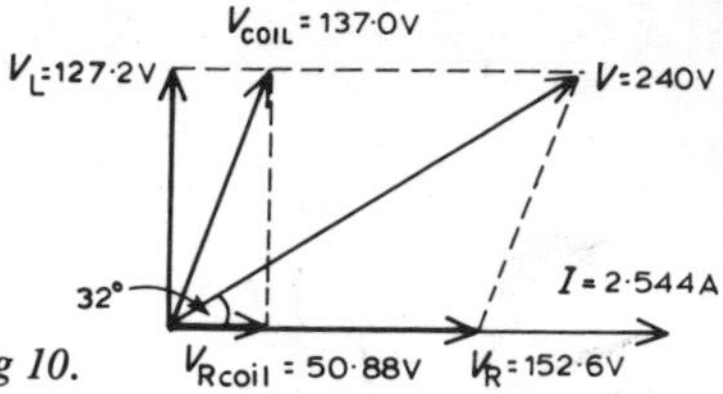

Fig 10

Problem 4 A coil of resistance 45 Ω and inductance 105.3 mH has an alternating voltage, $v = 141.4 \sin 628.4\,t$ volts applied across it. Calculate (a) the inductive reactance, (b) the circuit impedance, (c) the current flowing, (d) the p.d. across the resistance, (e) the p.d. across the inductance and (f) the phase angle between the supply voltage and current.

Since $v = 141.4 \sin 628.4\,t$ then $V_{MAX} = 141.4$ volts and $\omega = 2\pi f = 628.4$ rads/s. The r.m.s. voltage, $V = 0.707 \times 141.4 = 100$ V

and the supply frequency, $f = \frac{628.4}{2\pi} = 100$ Hz.

(a) Inductive reactance, $X_L = 2\pi fL = 2\pi(100)(105.3 \times 10^{-3})$ = **66.16 Ω**

(b) Circuit impedance $Z = \sqrt{(R^2 + X_L^2)} = \sqrt{(45^2 + 66.16^2)}$ = **80 Ω**

(c) Current $I = \frac{V}{Z} = \frac{100}{80}$ = **1.25 A**

(d) P.d. across the resistance, $V_R = IR = (1.25)(45)$ = **56.25 V**

(e) P.d. across the inductance, $V_L = IX_L = (1.25)(66.16)$ = **82.7 V**

(f) Phase angle between supply voltage and current, $\phi = \arctan \frac{X_L}{R}$

$= \arctan \frac{66.16}{45}$ = **55° 47′ lagging**

Problem 5 A 25 V, 1 kHz supply is connected across a coil having an inductance of 0.60 mH and resistance 2 Ω. Determine the supply current and its phase angle, and the active and reactive components of the current, showing each on a phasor diagram.

Inductive reactance $X_L = 2\pi fL$
$= 2\pi(1 \times 10^3)(0.60 \times 10^{-3}) = 3.77\ \Omega$
Impedance $Z = \sqrt{(R^2 + X_L^2)}$
$= \sqrt{(2^2 + 3.77^2)} = 4.268\ \Omega$
Supply current $I = \frac{V}{Z} = \frac{25}{4.268} = \mathbf{5.858\ A}$

Phase angle, $\phi = \arctan \frac{X_L}{R} = \arctan \frac{3.77}{2}$
$= \mathbf{62^\circ\ 3'}$ **lagging**
Active component of current $= I \cos\phi$,
where $\cos\phi = \frac{R}{Z} = (5.858)\left(\frac{2}{4.268}\right) = \mathbf{2.745\ A}$

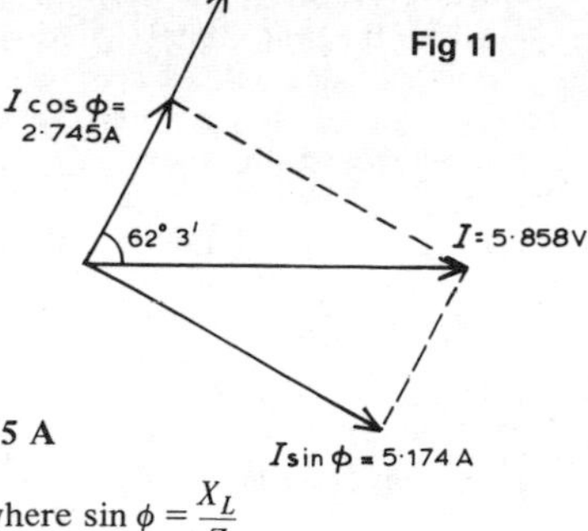

Reactive component of current $= I \sin\phi$, where $\sin\phi = \frac{X_L}{Z}$

$$= (5.858)\left(\frac{3.77}{4.268}\right) = \mathbf{5.174\ A}$$

The phasor diagram is shown in *Fig 11* where the active (or 'in-phase' component) is 'in-phase' with the supply voltage V.

R-C A.C. circuits

Problem 6 A resistance of 40 Ω is connected in series with a 50 μF capacitor. If the combination is connected across a 240 V, 50 Hz, supply, calculate (a) the impedance, (b) the current and (c) the phase angle between the supply voltage and the current.

$R = 40\ \Omega$; $C = 50\ \mu F = 50 \times 10^{-6}$ F; $V = 240$ V; $f = 50$ Hz.
The circuit diagram is as shown in *Fig 2(b)*.

Capacitive reactance $X_C = \frac{1}{2\pi fC} = \frac{1}{2\pi(50)(50 \times 10^{-6})} = 63.66\ \Omega$

(a) Impedance $Z = \sqrt{(R^2 + X_C^2)} = \sqrt{(40^2 + 63.66^2)} = \mathbf{75.18\ \Omega}$

(b) Current $I = \frac{V}{Z} = \frac{240}{75.18} = \mathbf{3.192\ A}$

(c) From *Fig 2(b)*, phase angle $\alpha = \arctan \frac{X_C}{R} = \arctan \frac{63.66}{40} = \mathbf{57^\circ\ 51'}$ **leading.**

('Leading' infers that the current is 'ahead' of the voltage and is in the position shown on the phasor diagram, since phasors revolve anticlockwise.)

Problem 7 A capacitor C is connected in series with a 10 Ω resistor across a supply of frequency 100 Hz. A current of 2 A flows and the circuit impedance is 26 Ω. Calculate (a) the value of capacitor C, (b) the supply voltage, (c) the circuit phase angle, (d) the p.d. across the resistor, and (e) the p.d. across the capacitor. Draw the phasor diagram.

(a) Impedance $Z = \sqrt{(R^2 + X_C^2)}$. Hence $X_C = \sqrt{(Z^2 - R^2)} = \sqrt{(26^2 - 10^2)} = 24\ \Omega$

$X_C = \dfrac{1}{2\pi fC}$. Hence capacitance $C = \dfrac{1}{2\pi f X_C} = \dfrac{1}{2\pi(100)(24)}$ F $= \mathbf{66.31\ \mu F}$

(b) Since $Z = \dfrac{V}{I}$, then $V = IZ = (2)(26) = \mathbf{52\ V}$

(c) Circuit phase angle $\alpha = \arctan \dfrac{X_C}{R}$

$= \arctan \dfrac{24}{10} = \mathbf{67^\circ\ 23'\ leading}$

(d) P.d. across R, $V_R = IR = (2)(10) = \mathbf{20\ V}$

(e) P.d. across C, $V_C = IX_C = (2)(24) = \mathbf{48\ V}$

The phasor diagram is shown in *Fig 12*, where the supply voltage V is the phasor sum of V_R and V_C.

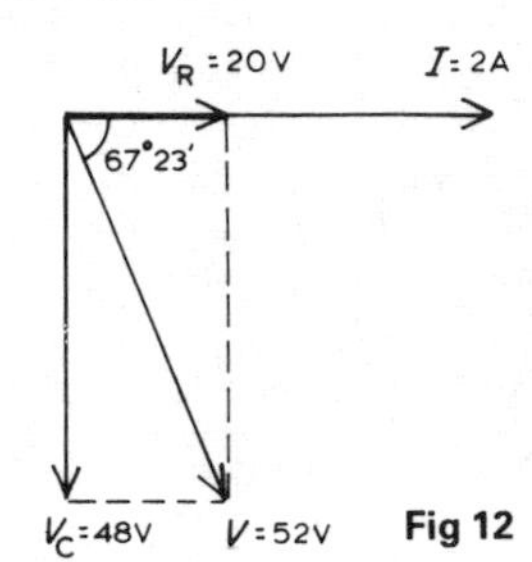

Fig 12

L-C-R A.C. circuits, series resonance and Q-factor

Problem 8 A coil of resistance 75 Ω and inductance 150 mH in series with an 8 μF capacitor, is connected to a 500 V, 200 Hz supply. Calculate (a) the current flowing, (b) the phase difference between the supply voltage and current, (c) the voltage across the coil and (d) the voltage across the capacitor. Sketch the phasor diagram.

The circuit diagram is shown in *Fig 3*, page 38.

Inductive reactance, $X_L = 2\pi fL = 2\pi(200)(150 \times 10^{-3}) = 188.5\ \Omega$

Capacitive reactance, $X_C = \dfrac{1}{2\pi fC} = \dfrac{1}{2\pi(200)(8 \times 10^{-6})} = 99.47\ \Omega$

Since $X_L > X_C$ the circuit is inductive (see phasor diagram in *Fig 3(b)*)

$X_L - X_C = 188.5 - 99.47 = 89.03\ \Omega$

Impedance $Z = \sqrt{[R^2 + (X_L - X_C)^2]} = \sqrt{[(75)^2 + (89.03)^2]} = 116.4\ \Omega$

(a) Current $I = \dfrac{V}{Z} = \dfrac{500}{116.4} = \mathbf{4.296\ A}$

(b) From *Fig 3(b)*, phase angle $\phi = \arctan\left(\dfrac{X_L - X_C}{R}\right) = \arctan\left(\dfrac{89.03}{75}\right)$

$= \mathbf{49^\circ\ 53'\ lagging}$

(c) Impedance of coil, $Z_{COIL} = \sqrt{(R^2 + X_L^2)} = \sqrt{[75^2 + 188.5^2]} = 202.9\ \Omega$

Voltage across coil, $V_{COIL} = IZ_{COIL} = (4.296)(202.9) = \mathbf{871.7\ V}$

Phase angle of coil, $\theta = \arctan \dfrac{X_L}{R} = \arctan \dfrac{188.5}{75} = 68^\circ\ 18'$ lagging

(d) Voltage across capacitor, $V_C = IX_C$
$= (4.296)(99.47) = \mathbf{427.3\ V}$
The phasor diagram is shown in *Fig 13*.
The supply voltage V is the phasor sum of V_{COIL} and V_C.

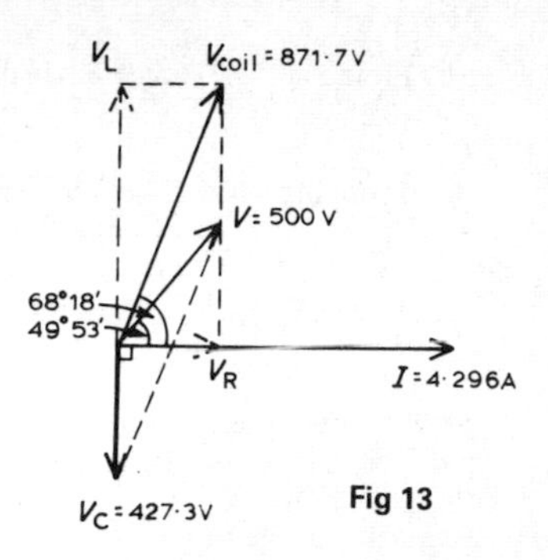

Fig 13

Problem 9 The following three impedances are connected in series across a 40 V, 20 kHz supply: (i) a resistance of 8 Ω, (ii) a coil of inductance 130 μH and 5 Ω resistance and (iii) a 10 Ω resistor in series with a 0.25 μF capacitor. Calculate (a) the circuit current, (b) the circuit phase angle and (c) the voltage drops across each impedance.

The circuit diagram is shown in *Fig 14(a)*. Since the total circuit resistance is 8 + 5 + 10, i.e. 23 Ω, an equivalent circuit diagram may be drawn as shown in *Fig 14(b)*.

Inductive reactance, $X_L = 2\pi fL = 2\pi(20 \times 10^3)(130 \times 10^{-6}) = 16.34\ \Omega$

Capacitive reactance, $X_C = \dfrac{1}{2\pi fC} = \dfrac{1}{2\pi(20 \times 10^3)(0.25 \times 10^{-6})} = 31.83\ \Omega$

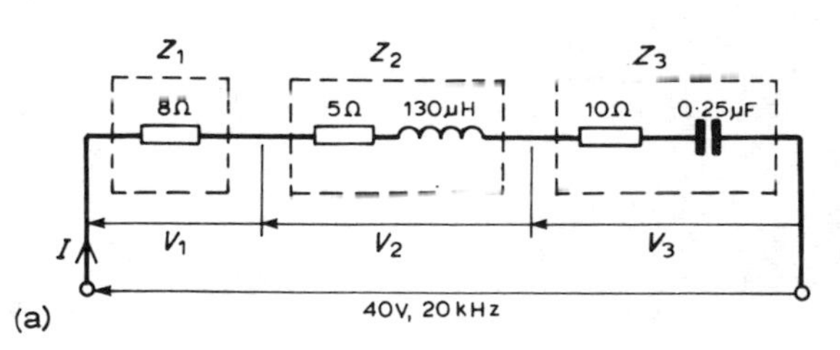

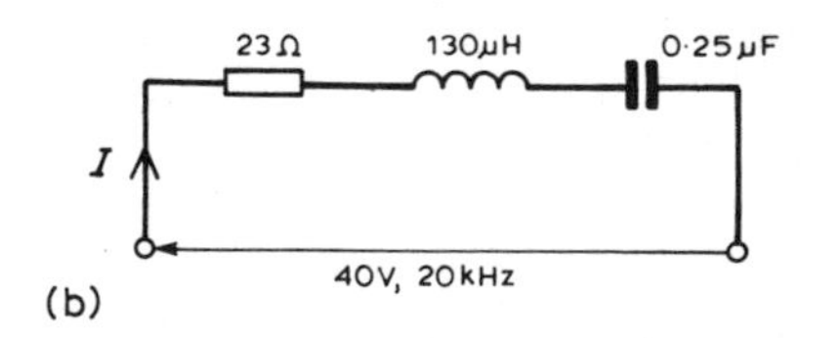

Fig 14

Since $X_C > X_L$ the circuit is capacitive (see phasor diagram in *Fig 3(c)*).
$X_C - X_L = 31.83 - 16.34 = 15.49\ \Omega$

(a) Circuit impedance, $Z = \sqrt{[R^2 + (X_C - X_L)^2]} = \sqrt{[23^2 + 15.49^2]} = 27.73\ \Omega$

Circuit current, $I = \dfrac{V}{Z} = \dfrac{40}{27.73} = \mathbf{1.442\ A}$

(b) From *Fig 3(c)*, circuit phase angle $\phi = \arctan\left(\frac{X_C - X_L}{R}\right) = \arctan\left(\frac{15.49}{23}\right)$

$= \mathbf{33^\circ\ 58'\ leading}$

(c) From *Fig 14(a)*. $V_1 = IR_1 = (1.442)(8) = \mathbf{11.54\ V}$

$V_2 = IZ_2 = I\sqrt{(5^2 + 16.34^2)} = (1.442)(17.09) = \mathbf{24.64\ V}$

$V_3 = IZ_3 = I\sqrt{(10^2 + 31.83^2)} = (1.442)(33.36) = \mathbf{48.11\ V}$

The 40 V supply voltage is the phasor sum of V_1, V_2 and V_3

Problem 10 Determine the p.d.'s V_1 and V_2 for the circuit shown in *Fig 15* if the frequency of the supply is 5 kHz. Draw the phasor diagram and hence determine the supply voltage V and the circuit phase angle.

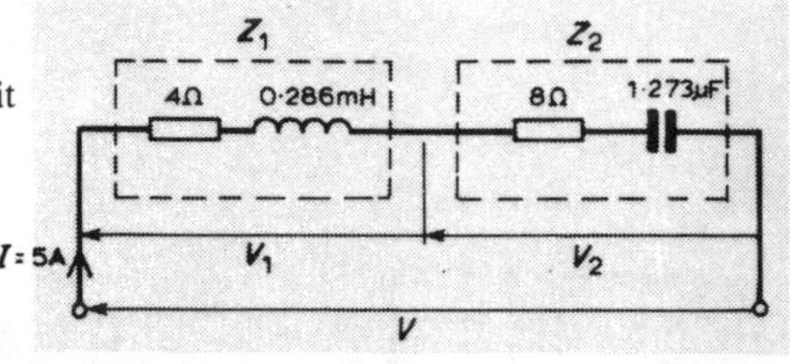

Fig 15

For impedance Z_1: $R_1 = 4\ \Omega$ and $X_L = 2\pi fL = 2\pi(5 \times 10^3)(0.286 \times 10^{-3})$

$= 8.985\ \Omega$

$V_1 = IZ_1 = I\sqrt{(R^2 + X_L^2)} = 5\sqrt{(4^2 + 8.985^2)} = 49.18\ V$

Phase angle $\phi_1 = \arctan\left(\frac{X_L}{R}\right) = \arctan\left(\frac{8.985}{4}\right) = \mathbf{66^\circ\ 0'\ lagging}$

For impedance Z_2: $R_2 = 8\ \Omega$ and $X_C = \frac{1}{2\pi fC} = \frac{1}{2\pi(5 \times 10^3)(1.273 \times 10^{-6})}$

$= 25.0\ \Omega$

$V_2 = IZ_2 = I\sqrt{(R^2 + X_C^2)} = 5\sqrt{(8^2 + 25.0^2)} = 131.2\ V$

Phase angle $\phi_2 = \arctan\left(\frac{X_C}{R}\right) = \arctan\left(\frac{25.0}{8}\right) = 72^\circ\ 15'$ leading.

The phasor diagram is shown in *Fig 16*. The phasor sum of V_1 and V_2 gives the supply voltage V of **100 V** at a phase angle of **53° 8′ leading.** These values may be determine by drawing or by calculation —either by resolving into horizontal and vertical components or by the cosine and sine rules.

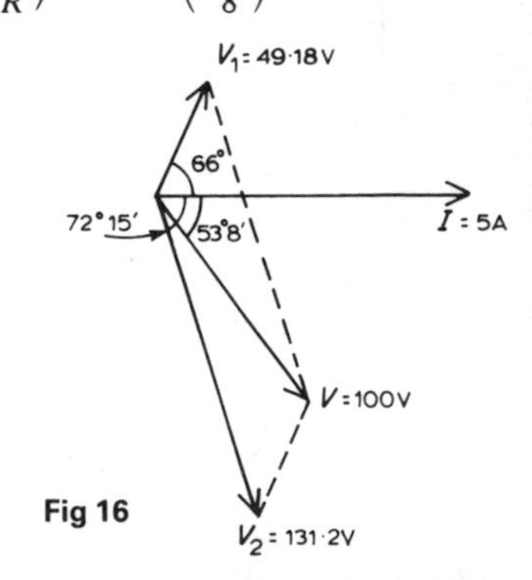

Fig 16

Problem 11 A coil having a resistance of 20 Ω and inductance 80 mH is connected in series with a 50 μF capacitor across a 150 V supply. At what frequency does resonance occur? Calculate the current flowing at the resonant frequency.

$$\text{Resonant frequency } f_r = \frac{1}{2\pi\sqrt{(LC)}} = \frac{1}{2\pi\sqrt{\left(\frac{80}{10^3}\right)\left(\frac{50}{10^6}\right)}} = \frac{1}{2\pi\sqrt{\left(\frac{80 \times 5}{10^8}\right)}}$$

$$= \frac{1}{2\pi\left(\frac{\sqrt{400}}{10^4}\right)} = \frac{10^4}{2\pi(20)} = \mathbf{79.58\ Hz}$$

At resonance, impedance $Z = R$

Hence current at resonance $I = \frac{V}{R} = \frac{150}{20} = \mathbf{7.5\ A}$

Problem 12 The current at resonance in a series *L-C-R* circuit is 200 μA. If the applied voltage is 5 mV at a frequency of 100 kHz, and the circuit inductance is 50 mH, find (a) the circuit resistance and (b) the circuit capacitance.

(a) Current $I = 200\ \mu\text{A} = 200 \times 10^{-6}$ A; Voltage $V = 5\text{ mV} = 5 \times 10^{-3}$ V
At resonance, impedance $Z = R$.

$$\text{Hence resistance } R = \frac{V}{I} = \frac{5 \times 10^{-3}}{200 \times 10^{-6}} = \frac{5 \times 10^6}{200 \times 10^3} = \mathbf{25\ \Omega}$$

(b) At resonance $X_L = X_C$, i.e. $2\pi f_r L = \frac{1}{2\pi f_r C}$

$$\text{Hence capacitance } C = \frac{1}{(2\pi f_r)^2 L} = \frac{1}{(2\pi \times 100 \times 10^3)^2(50 \times 10^{-3})}\text{ F}$$

$$= \frac{(10^{12})(10^3)}{4\pi^2(10^{10})(50)}\text{ pF} = \mathbf{50.7\ pF}$$

Problem 13 A series circuit comprises a coil of resistance 2 Ω and inductance 60 mH, and a 30 μF capacitor. Determine the Q-factor of the circuit at resonance.

$$\text{At resonance, Q-factor} = \frac{1}{R}\sqrt{\left(\frac{L}{C}\right)} = \frac{1}{2}\sqrt{\left(\frac{60 \times 10^{-3}}{30 \times 10^{-6}}\right)} = \frac{1}{2}\sqrt{\left(\frac{60 \times 10^6}{30 \times 10^3}\right)}$$

$$= \frac{1}{2}\sqrt{(2000)} = \mathbf{22.36}$$

Problem 14 A coil of negligible resistance and inductance 100 mH is connected in series with a capacitance of 2 μF and a resistance of 10 Ω across a 50 V, variable frequency supply. Determine (a) the resonant frequency, (b) the current at resonance, (c) the voltages across the coil and the capacitor at resonance, and (d) the Q-factor of the circuit.

$$\text{(a) Resonant frequency, } f_r = \frac{1}{2\pi\sqrt{(LC)}} = \frac{1}{2\pi\sqrt{\left(\frac{100}{10^3} \times \frac{2}{10^6}\right)}} = \frac{1}{2\pi\sqrt{\left(\frac{20}{10^8}\right)}}$$

$$= \frac{1}{\frac{2\pi\sqrt{20}}{10^4}} = \frac{10^4}{2\pi\sqrt{20}} = \mathbf{355.9\ Hz}$$

(b) Current at resonance $I = \frac{V}{R} = \frac{50}{10} = \mathbf{5\ A}$

(c) Voltage across coil at resonance, $V_L = IX_L = I(2\pi f_r L)$
$= (5)(2\pi \times 355.9 \times 100 \times 10^{-3}) = \mathbf{1118\ V}$

Voltage across capacitance at resonance, $V_C = IX_C = \frac{1}{2\pi f_r C}$
$= \frac{5}{2\pi(355.9)(2 \times 10^{-6})}$
$= \mathbf{1118\ V}$

(d) Q-factor (i.e. voltage magnification at resonance) $= \frac{V_L}{V}$ or $\frac{V_C}{V}$
$= \frac{1118}{50} = \mathbf{22.36}$

[Q-factor may also have been determined by $\frac{2\pi f_r L}{R}$ or $\frac{1}{2\pi f_r CR}$

or $\frac{1}{R}\sqrt{\left(\frac{L}{C}\right)}$ (see para. 15)

Problem 15 A filter in the form of a series *L-R-C* circuit is designed to operate at a resonant frequency of 5 kHz. Included within the filter is a 20 mH inductance and 10 Ω resistance. Determine the bandwidth of the filter.

Q-factor at resonance is given by

$$Q_r = \frac{\omega_r L}{R} = \frac{(2\,\pi\,5000)(20 \times 10^{-3})}{10} = 62.83$$

Since $Q_r = f_r/(f_2 - f_1)$,

bandwidth, $(f_2 - f_1) = \frac{f_r}{Q_r} = \frac{5000}{62.83} = \mathbf{79.6\ Hz}$

POWER IN A.C. CIRCUITS

Problem 16 An instantaneous current $i = 150 \sin \omega t$ mA flows through a pure resistance of 4 kΩ. Determine the power dissipated in the resistor.

Power dissipated, $P = I^2 R$, where I is the r.m.s. value of current.
If $i = 150 \sin \omega t$ mA, then $I_{MAX} = 150$ mA and r.m.s. current, $I = 0.707 \times 150$ $= 106.1$ mA $= 0.1061$ A
Hence power $P = I^2 R = (0.1061)^2(4000) = \mathbf{45.03\ W}$

Problem 17 A coil of inductance 40 mH and resistance 30 Ω is connected to a 200 V, 80 Hz supply. Calculate the power dissipated.

Inductive reactance, $X_L = 2\pi fL = 2\pi(80)(40 \times 10^{-3}) = 20.11\ \Omega$
Impedance $Z = \sqrt{(R^2 + X_L^2)} = \sqrt{(30^2 + 20.11^2)} = 36.12\ \Omega$

Current $I = \dfrac{V}{Z} = \dfrac{200}{36.12} = 5.537$ A

To calculate power dissipated in an a.c. circuit either of two formulae may be used.
(i) $P = I^2R = (5.537)^2(30) =$ **919.8 W**, or

(ii) $P = VI \cos\phi$, where $\cos\phi = \dfrac{R}{Z} = \dfrac{30}{36.12} = 0.8306$

Hence $P = (200)(5.537)(0.8306) =$ **919.8 W**

Problem 18 A pure inductance is connected to a 100 V, 50 Hz supply and the apparent power of the circuit is 250 VA. Find the value of the inductance.

Apparent power, $S = VI$. Hence current $I = \dfrac{S}{V} = \dfrac{250}{100} = 2.5$ A

Inductive reactance, $X_L = \dfrac{V}{I} = \dfrac{100}{2.5} = 40\ \Omega$

Since $X_L = 2\pi fL$, inductance $L = \dfrac{X_L}{2\pi f} = \dfrac{40}{2\pi 50} =$ **0.127 H or 127 mH**

Problem 19 A transformer has a rated output of 300 kVA at a power factor of 0.8. Determine the rated power output and the corresponding reactive power.

$VI = 300$ kV A $= 300 \times 10^3$ V A; p.f. $= 0.8 = \cos\phi$
Power output, $P = VI \cos\phi = (300 \times 10^3)(0.8) =$ **240 kW**
Reactive power $Q = VI \sin\phi$
If $\cos\phi = 0.8$, then $\phi = \arccos 0.8 = 36°\ 52'$
Hence $\sin\phi = \sin 36°\ 52' = 0.6$
Reactive power, $Q = (300 \times 10^3)(0.6) =$ 180 kvar

Problem 20 A load takes 150 kW at a power factor of 0.6 lagging. Calculate the apparent power and the reactive power.

True power $P = 150$ kW $= VI \cos\phi$. Power factor $= 0.6 = \cos\phi$

Apparent power, $S = VI = \dfrac{150}{0.6} =$ **kVA.**

Angle $\phi = \arccos 0.6 = 53°\ 8'$. Hence $\sin\phi = \sin 53°\ 8' = 0.8$
Hence reactive power, $Q = VI \sin\phi = (250 \times 10^3)(0.8) =$ 200 kvar

Problem 21 The power taken by an inductive circuit when connected to a 240 V, 50 Hz supply is 800 W and the current is 4 A. Calculate (a) the resistance, (b) the impedance, (c) the reactance, (d) the inductance, (e) the power factor and (f) the phase angle between voltage and current.

(a) Power, $P = I^2R$. Hence $R = \dfrac{P}{I^2} = \dfrac{800}{4^2} =$ **50 Ω**

(b) Impedance, $Z = \dfrac{V}{I} = \dfrac{240}{4} =$ **60 Ω**

(c) Since $Z = \sqrt{(R^2 + X_L^2)}$ then $X_L = \sqrt{(Z^2 - R^2)} = \sqrt{(60^2 - 50^2)} =$ **33.17 Ω**

(d) Inductive reactance, $X_L = 2\pi fL$. Hence $L = \frac{X_L}{2\pi f} = \frac{33.17}{2\pi 50}$

$= \mathbf{0.1056\ H\ or\ 105.6\ mH}$

(e) Power factor $= \frac{\text{true power}}{\text{apparent power}} = \frac{VI \cos \phi}{VI} = \frac{800}{(240)(4)} = \mathbf{0.8333}$

(f) p.f. $= \cos \phi = 0.8333$. Hence phase angle, $\phi = \arccos 0.8333$
$= \mathbf{33^\circ\ 34'\ lagging}$

Problem 22 A series circuit comprising a capacitor and a resistor takes 225 W at a power factor of 0.75 from a 120 V, 100 Hz supply. Determine (a) the current flowing, (b) the phase angle, (c) the resistance, (d) the impedance and (e) the capacitance.

(a) Power factor $= \frac{\text{true power}}{\text{apparent power}}$, i.e. $0.75 = \frac{225}{(120)(I)}$

Hence current $I = \frac{225}{(120)(0.75)} = \mathbf{2.5\ A}$

(b) Power factor $= 0.75 = \cos \phi$. Hence phase angle, $\phi = \arccos 0.75$
$= \mathbf{41^\circ\ 25'\ leading}$

(c) Power, $P = I^2 R$. Hence $R = \frac{P}{I^2} = \frac{225}{(2.5)^2} = \mathbf{36\ \Omega}$

(d) Impedance $Z = \frac{V}{I} = \frac{120}{2.5} = \mathbf{48\ \Omega}$

(e) Capacitive reactance, $X_C = \sqrt{(Z^2 - R^2)} = \sqrt{(48^2 - 36^2)} = 31.75\ \Omega$

$X_C = \frac{1}{2\pi fC}$. Hence capacitance, $C = \frac{1}{2\pi f X_C} = \frac{1}{2\pi(100)(31.75)}$ F

$= \mathbf{50.13\ \mu F}$

Problem 23 A single phase motor is connected to a 400 V, 50 Hz supply. The motor develops 15 kW with an efficiency of 80% and a power factor of 0.75 lagging. Determine (a) the input kVA's, (b) the active and reactive components of the current, and (c) the reactive voltamperes.

(a) Efficiency $= \frac{\text{output power}}{\text{input power}} = \frac{\text{output power}}{VI \cos \phi} = \frac{\text{output power}}{(VI)\ (\text{p.f.})}$

Hence $\frac{80}{100} = \frac{15 \times 10^3}{(VI)(0.75)}$, from which, $VI = \frac{15 \times 10^3}{(0.80)(0.75)} = \mathbf{25\ 000\ VA.}$

Therefore, input kilovoltamperes = **25 kV A**

(b) Current taken by motor $= \frac{\text{input voltamperes}}{\text{voltage}} = \frac{VI}{V} = \frac{25\ 000}{400}$

$= \mathbf{62.5\ A}$

Active or in-phase component of current $= I \cos \phi = (62.5)(0.75)$
$= \mathbf{46.88\ A}$

Since p.f. $= \cos \phi = 0.75$ then $\phi = \arccos 0.75 = 41.41^\circ$
Hence $\sin \phi = \sin 41.41^\circ = 0.661\ 4$

Reactive or quadrative component of current = $I \sin \phi$ = (62.5)(0.661 4)
= **41.34 A**

(c) Reactive voltamperes = $VI \sin \phi$ = (400)(41.34) = **16.54 kvar**

C. FURTHER PROBLEMS ON SERIES A.C. CIRCUITS

SHORT ANSWER PROBLEMS

1 Draw phasor diagrams to represent (a) a purely resistive a.c. circuit, (b) a purely inductive a.c. circuit, and (c) a purely capacitive a.c. circuit.
2 What is inductive reactance? State the symbol, the unit and the formula for determining inductive reactance.
3 What is capacitive reactance? State the symbol, the unit and the formula for determining capacitive reactance.
4 Draw phasor diagrams to represent (a) a coil (having both inductance and resistance), and (b) a series capacitive circuit containing resistance.
5 What does 'impedance' mean when referring to an a.c. circuit?
6 Draw and 'impedance triangle' for an *R-C* circuit. From the triangle derive an expression for (a) impedance, and (b) phase angle.
7 State two formulae which may be used to calculate power in a series a.c. circuit.
8 What is series resonance?
9 Derive a formula for resonant frequency f_r in terms of L and C.
10 What does the Q-factor of a series circuit mean?
11 State three formulae used to calculate the Q-factor of a series circuit at resonance.
12 State an advantage of a high Q-factor in a series high-frequency circuit.
13 State a disadvantage of a high Q-factor in a series power circuit.
14 Define 'power factor'.
15 Define (a) apparent power, (b) reactive power.
16 In an a.c. circuit (a) the horizontal component of current is called the ....
and (b) the vertical component of current is called the ....
17 Define (a) bandwidth, (b) selectivity

MULTI-CHOICE PROBLEMS (answers on page 191)

1 An inductance of 100 mH connected across a 100 V, 50 Hz supply has an inductive reactance of (a) 10π kΩ; (b) 10π Ω; (c) 1000π Ω; (d) π Ω.
2 When the frequency of an a.c. circuit containing resistance and inductance is increased, the current (a) decreases; (b) increases; (c) stays the same.
3 In *Problem 2*, the phase angle of the circuit (a) decreases; (b) increases; (c) stays the same.
4 When the frequency of an a.c. circuit containing resistance and capacitance is decreased, the current (a) increases; (b) decreases; (c) stays the same.
5 In *Problem 4*, the phase angle of the circuit (a) increases; (b) decreases; (c) stays the same.

6 A 10 μF capacitor is connected to a 5 kHz supply. The capacitive reactance is (a) $\frac{10}{\pi}$ kΩ; (b) $\frac{\pi}{10}$ Ω; (c) $\frac{10}{\pi}$ Ω; (d) $\frac{10}{\pi}$ mΩ.

7 In a series a.c. circuit the voltage across a pure inductance is 15 V and the voltage across a pure resistance is 8 V. The supply voltage is (a) 17 V; (b) 23 V; (c) 7 V; (d) 1.875 V.

8 The impedance of a coil, which has a resistance of A ohms and an inductance of B henry's, connected across a supply of frequency C Hz is (a) $2\pi BC$; (b) $A+B$; (c) $\sqrt{(A^2+B^2)}$; (d) $\sqrt{\{A^2+(2\pi BC)^2\}}$.

9 In *Problem 8*, the phase angle between the current and the applied voltage is given by: (a) $\arctan\left(\frac{B}{A}\right)$; (b) $\arctan\left(\frac{2\pi BC}{A}\right)$; (c) $\arctan\left(\frac{A}{2\pi BC}\right)$; (d) $\tan\left(\frac{2\pi BC}{A}\right)$

10 Which of the following statements is false?
(a) The product of r.m.s. current and voltage gives the apparent power in an a.c. circuit.
(b) Impedance is at a minimum at resonance in an a.c. circuit.
(c) Current is at a maximum at resonance in an a.c. circuit.
(d) $\frac{\text{Apparent power}}{\text{true power}}$ = power factor.

11 In an R-L-C a.c. series circuit a current of 2 A flows when the supply voltage is 100 V. The phase angle between current and voltage is 60° leading. Which of the following statements is false?
(a) The circuit is effectively capacitive.
(b) The apparent power is 200 V A.
(c) The equivalent circuit reactance is 50 Ω.
(d) The true power is 100 W.

12 A series a.c. circuit comprising a coil of inductance 100 mH and resistance 1 Ω and a 10 μF capacitor is connected across a 10 V supply. At resonance the p.d. across the capacitor is (a) 10 kV; (b) 1 kV; (c) 100 V; (d) 10 V.

CONVENTIONAL PROBLEMS

R-L A.C. circuits

1 A coil has a resistance of 6 Ω and an inductance of 15 mH and is connected to a 25 V, 50 Hz supply. Calculate (a) the inductive reactance; (b) the impedance; (c) the current and (d) the phase angle between the voltage and current.
[(a) 4.71 Ω; (b) 7.63 Ω; (c) 3.28 A; (d) 38°8′ lagging]

2 A coil takes a current of 3 A from a 24 V d.c. supply. When connected to a 150 V, 50 Hz a.c. supply the current is 10 A. Calculate (a) the resistance; (b) the impedance and (c) the inductance of the coil.
[(a) 8 Ω; (b) 15 Ω; (c) 40.39 mH]

3 A pure inductance of 3.24 mH is connected in series with a resistance of 20 Ω. If the frequency of the sinusoidal supply is 2 kHz and the p.d. across the resistance is 15 V, determine the value of the supply voltage and the voltage across the pure inductance. Draw the phasor diagram.
[34.02 V; 30.53 V]

4 An alternating voltage given by v = 100 sin 314.2 t volts is applied across a coil of inductance 100 mH and resistance 30 Ω. Determine (a) the circuit impedance;

(b) the current; (c) the p.d. across the resistance; (d) the p.d. across the inductance, and (e) the circuit phase angle.

[(a) 43.44 Ω; (b) 1.628 A; (c) 48.84 V; (d) 51.15 V; (e) 46°19′ lagging]

5 A 40 V, 2.5 kHz supply is connected across a coil having an inductance of 0.40 mH and resistance 3 Ω. Determine the supply current and its phase angle and the active and reactive components of the current, showing each on a phasor diagram.

[5.745 A; 64°29′ lagging; 2.475 A; 5.185 A]

R-C A.C. circuits

6 A 60 μF capacitor and a resistance of 40 Ω are connected in series across a 250 V, 50 Hz supply. Calculate (a) the impedance; (b) the current; and (c) the phase angle between voltage and current.

[(a) 66.44 Ω; (b) 3.763 A; (c) 52°59′ leading]

7 A 15 Ω resistor is connected in series with an unknown capacitor C across a 200 Hz frequency supply. The circuit impedance is 30 Ω when a current of 3 A flows. Calculate (a) the value of capacitor C; (b) the supply voltage; (c) the circuit phase angle; (d) the p.d. across the resistor; and (e) the p.d. across the capacitor. Draw the phasor diagram.

[(a) 30.63 μF; (b) 90 V; (c) 60° leading; (d) 45 V; (e) 77.94 V]

8 A 32 μF capacitor and a 40 Ω resistor are connected in series across a 300 V supply. If the current flowing is 5 A find (a) the frequency of the supply; (b) the p.d. across the resistor and (c) the p.d. across the capacitor.

[(a) 111.2 Hz; (b) 200 V; (c) 223.6 V]

L-C-R a.c. circuits, series resonance and Q-factor

9 A coil of inductance 74 mH and resistance 28 Ω in series with a 50 μF capacitor is connected to a 250 V, 50 Hz supply. Calculate (a) the impedance; (b) the current flowing; (c) the phase difference between voltage and current; (d) the voltage across the coil; and (e) the voltage across the capacitor. Sketch the phasor diagram.

[(a) 49.16 Ω; (b) 5.085 A, (c) 55°17′ leading; (d) 185.1 V; (e) 323.7 V]

10 Three impedances are connected in series across a 100 V, 2 kHz supply. The impedances comprise:

(i) an inductance of 0.45 mH and 2 Ω resistance,
(ii) an inductance of 570 μH and 5 Ω resistance, and
(iii) a capacitor of 10 μF and 3 Ω resistance.

Assuming no mutual inductive effects between the two inductances calculate (a) the circuit impedance; (b) the circuit current; (c) the circuit phase angle and (d) the voltage across each impedance. Draw the phasor diagram.
[(a) 11.12 Ω; (b) 8.99 A; (c) 25°56′ lagging; (d) 53.92 V; 78.53 V; 76.46 V]

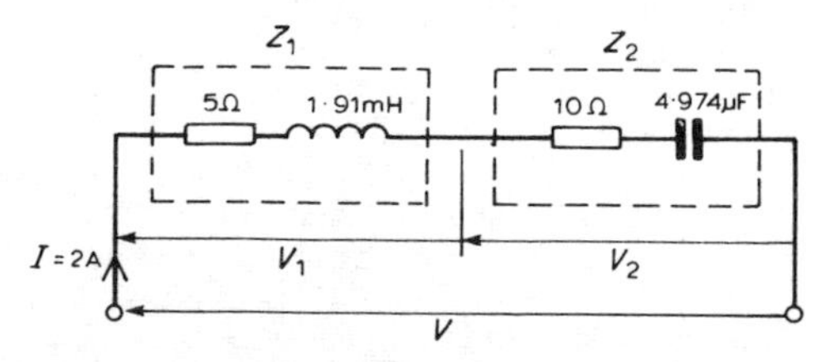

Fig 17

11 For the circuit shown in *Fig 17* determine the voltages V_1 and V_2 if the supply

frequency is 1 kHz. Draw the phasor diagram and hence determine the supply voltage V and the circuit phase angle.

[V_1 = 26.0 V; V_2 = 67.05 V; V = 50 V; 53°8′ leading]

12 A coil having an inductance of 50 mH and resistance 10 Ω is connected in series with a 40 μF capacitor across a 100 V supply. At what frequency does resonance occur? Calculate the current flowing at the resonant frequency.

[112.5 Hz; 10 A]

13 The current at resonance in a series L-C-R circuit is 5 mA. If the applied voltage is 300 mV at a frequency of 50 kHz and the circuit capacitance is 0.03 μF, find the circuit resistance and inductance.

[60 Ω; 337.7 μH]

14 A series circuit comprises a coil of resistance 20 Ω and inductance 2 mH and a 500 pF capacitor. Determine the Q-factor of the circuit at resonance. If the supply voltage is 1.5 V, what is the voltage across the capacitor?

[100; 150 V]

15 Calculate the inductance of a coil which must be connected in series with a 2000 pF capacitor to give a resonant frequency of 250 kHz.

[202.6 μH]

16 A coil of inductance 50 mH and negligible resistance is connected in series with a 5 μF capacitor and a resistance of 12.5 Ω across a 200 V, variable frequency supply. Calculate (a) the resonant frequency; (b) the current at resonance; (c) the voltage across the coil and the capacitor at resonance; and (d) the Q-factor of the circuit.

[(a) 318.3 Hz; (b) 16 A; (c) 1600 V; (d) 8]

17 A coil of inductance 0.2 H and 10 Ω resistance is connected in series with a capacitor across a 150 V, 100 Hz supply. If the current is in phase with the supply voltage, determine the value of the capacitance and the p.d. across its terminals.

[12.67 μF; 1884 V]

Power in a.c. circuits

18 A voltage $v = 300 \sin \omega t$ volts is applied across a resistance of 10 kΩ. Find the power dissipated in the resistor.

[4.50 W]

19 A 10 μF capacitor is connected to a 200 V, 100 Hz supply. Determine the true power and the apparent power.

[0; 251.3 V A]

20 A coil of inductance 5 mH and resistance 20 Ω is connected to a 40 V, 1 kHz supply. Calculate the power dissipated.

[23.07 W]

21 A pure inductance is connected to a 120 V, 60 Hz supply and the apparent power of the circuit is 200 VA. Find the value of the inductance.

[191 mH]

22 A motor takes a current of 12 A when supplied from a 240 V a.c. supply. Assuming a power factor of 0.75 lagging find the power consumed.

[2.16 kW]

23 A transformer has a rated output of 200 kVA at a power factor of 0.7. Determine the rated power output and the corresponding reactive power.

[140 kW; 142.8 kvar]

24 The reactive load on a substation is 120 kvar and the apparent power supplied is 150 kVA. Calculate the corresponding power and power factor.

[90 kW; 0.6]

25 A load takes 90 kW at a power factor of 0.75 lagging. Calculate the apparent power and the reactive power.

[120 kVA; 79.37 kvar]

26 The power taken by an inductive circuit when connected to a 200 V, 50 Hz supply is 1.2 kW and the current is 8 A. Calculate (a) the resistance; (b) the impedance; (c) the reactance; (d) the inductance; (e) the power factor and (f) the circuit phase angle.

[(a) 18.75 Ω; (b) 25 Ω; (c) 16.54 Ω; (d) 52.65 mH; (e) 0.75; (f) 41°25′ lagging]

27 A capacitor connected in series with a resistor takes 300 W at a power factor of 0.80 from a 150 V, 200 Hz supply. Calculate (a) the current flowing; (b) the circuit phase angle; (c) the resistance; (d) the impedance, and (e) the capacitance.

[(a) 2.5 A; (b) 36°52′ leading; (c) 48 Ω; (d) 60 Ω; (e) 22.10 μF]

28 A single phase motor is connected to a 415 V, 50 Hz supply. The motor develops 12 kW at an efficiency of 75% and a power factor of 0.8 lagging. Determine (a) the input kVA; (b) the active and reactive components of the current; and (c) the reactive voltampers.

[(a) 20 kVA; (b) 38.55 A; 28.91 A; (c) 12 kvar]

3 Single-phase parallel a.c. circuits

A. MAIN POINTS CONCERNED WITH SINGLE-PHASE PARALLEL A.C. CIRCUITS

1 In parallel circuits, such as those shown in *Fig 1*, the voltage is common to each branch of the network and is thus taken as the reference phasor when drawing phasor diagrams.

2 ***R-L* parallel circuit.** In the two branch parallel circuit containing resistance R and inductance L shown in *Fig 1(a)*, the current flowing in the resistance, I_R, is in-phase with the supply voltage V and the current flowing in the inductance, I_L, lags the supply voltage by 90°. The supply current I is the phasor sum of I_R and I_L and thus the current I lags the applied voltage V by an angle lying between 0° and 90° (depending on the values of I_R and I_L), shown as angle ϕ in the phasor diagram.

From the phasor diagram: $I = \sqrt{(I_R^2 + I_L^2)}$, (by Pythagoras' theorem)

where $I_R = \dfrac{V}{R}$ and $I_L = \dfrac{V}{X_L}$

$$\tan\phi = \frac{I_L}{I_R},\ \sin\phi = \frac{I_L}{I} \text{ and } \cos\phi = \frac{I_R}{I}$$

(by trigonometric ratios)

$$\text{Circuit impedance, } Z = \frac{V}{I}$$

3 ***R-C* parallel circuit.** In the two branch parallel circuit containing resistance R and capacitance C shown in *Fig 1(b)*, I_R is in-phase with the supply voltage V and the current flowing in the capacitor, I_C, leads V by 90°. The supply current I is the phasor sum of I_R and I_C and thus the current I leads the applied voltage V by an angle lying between 0° and 90° (depending on the values of I_R and I_C), shown as angle α in the phasor diagram.

From the phasor diagram: $I = \sqrt{(I_R^2 + I_C^2)}$, (by Pythagoras' theorem)

where $I_R = \dfrac{V}{R}$ and $I_C = \dfrac{V}{X_C}$

$\tan\alpha = \dfrac{I_C}{I_R}$, $\sin\alpha = \dfrac{I_C}{I}$ and $\cos\alpha = \dfrac{I_R}{I}$ (by trigonometric ratios)

Circuit impedance $Z = \dfrac{V}{I}$

CIRCUIT DIAGRAM PHASOR DIAGRAM

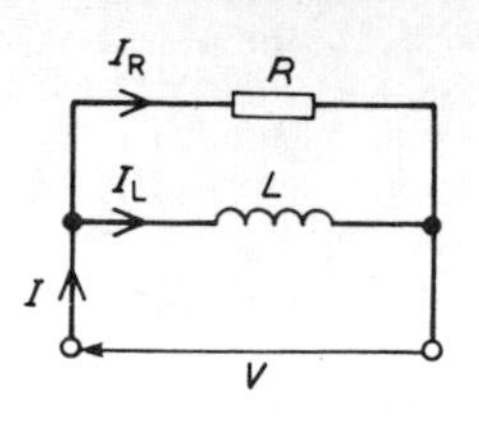

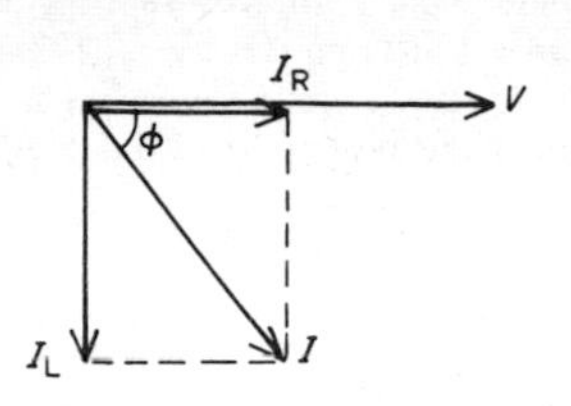

(a)

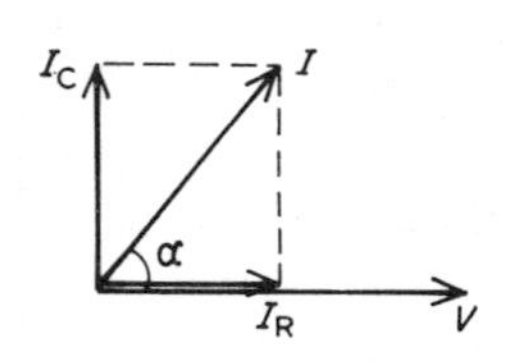

(b)

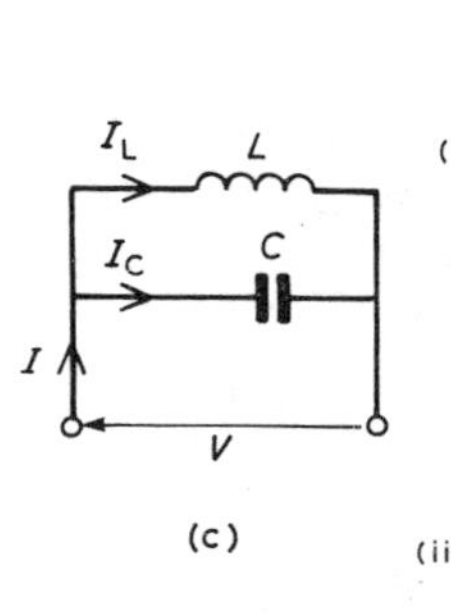

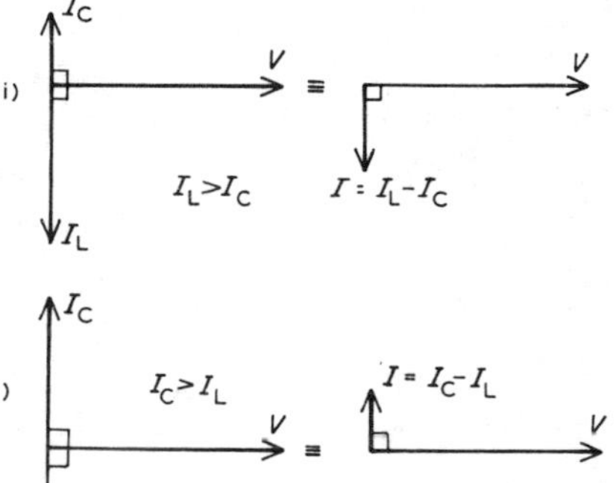

Fig 1

4 *L-C* **parallel circuit.** In the two branch parallel circuit containing inductance L and capacitance C shown in *Fig 1(c)*, I_L lags V by 90° and I_C leads V by 90°. Theoretically there are three phasor diagrams possible—each depending on the relative values of I_L and I_C:

(i) $I_L > I_C$ (giving a supply current, $I = I_L - I_C$ lagging V by 90°)
(ii) $I_C > I_L$ (giving a supply current, $I = I_C - I_L$ leading V by 90°)
(iii) $I_L = I_C$ (giving a supply current, $I = 0$).

The latter condition is not possible in practice due to circuit resistance inevitably being present (as in the circuit described in para. 5).

For the *L-C* parallel circuit, $I_L = \frac{V}{X_L}$, $I_C = \frac{V}{X_C}$

I = phasor difference between I_L and I_C, and $Z = \frac{V}{I}$.

5 *LR-C* **parallel circuit.** In the two branch circuit containing capacitance C in parallel with inductance L and resistance R in series (such as a coil) shown in *Fig 2(a)*, the phasor diagram for the LR branch alone is shown in *Fig 2(b)* and the phasor diagram for the C branch is shown alone in *Fig 2(c)*. Rotating each and superimposing on one another gives the complete phasor diagram shown in *Fig 2(d)*.

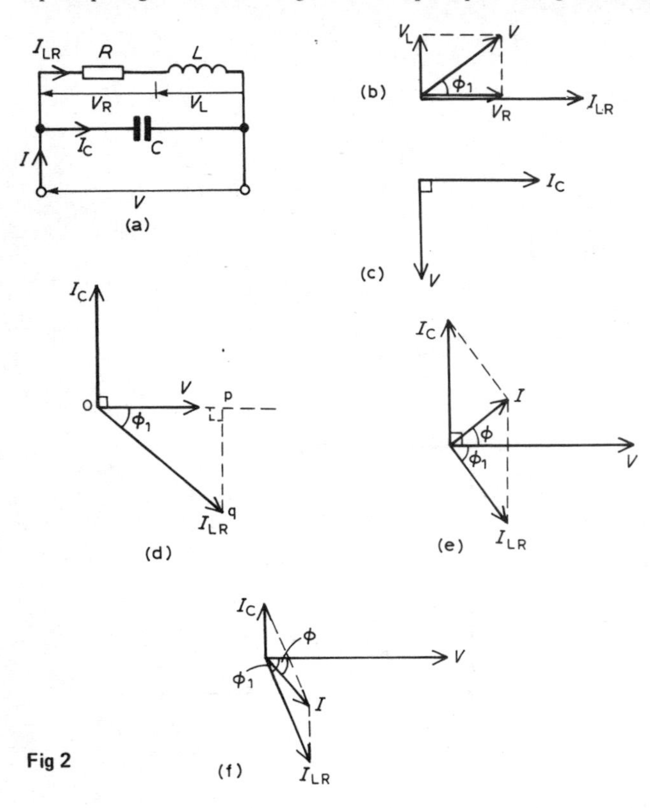

Fig 2

6 The current I_{LR} of *Fig 2(d)* may be resolved into horizontal and vertical components. The horizontal component, shown as op is $I_{LR}\cos\phi_1$, and the vertical component, shown as pq is $I_{LR}\sin\phi_1$. There are three possible conditions for this circuit:

(i) $I_C > I_{LR}\sin\phi_1$ (giving a supply current I leading V by angle ϕ—as shown in *Fig 2(e)*).

(ii) $I_{LR}\sin\phi > I_C$ (giving I lagging V by angle ϕ—as shown in *Fig 2(f)*).

(iii) $I_C = I_{LR}\sin\phi_1$ (this is called parallel resonance, see para. 10).

7 There are two methods of finding the phasor sum of currents I_{LR} and I_C in *Figs 2(e) and (f)*. These are: (i) by a scaled phasor diagram, or (ii) by resolving each current into their 'in-phase' (i.e. horizontal) and 'quadrature' (i.e. vertical) components (see chapter 2, para. 21).

8 With reference to the phasor diagrams of *Fig 2*:
Impedance of LR branch, $Z_{LR} = \sqrt{(R^2 + X_L^2)}$

Current, $I_{LR} = \dfrac{V}{Z_{LR}}$ and $I_C = \dfrac{V}{X_C}$

Supply current I = phasor sum of I_{LR} and I_C (by drawing)
$= \sqrt{\{(I_{LR} \cos \phi_1)^2 + (I_{LR} \sin \phi_1 \sim I_C)^2\}}$ (by calculation)
where ~ means 'the difference between'.

Circuit impedance $Z = \dfrac{V}{I}$

$\tan \phi_1 = \dfrac{V_L}{V_R} = \dfrac{X_L}{R}$, $\sin \phi_1 = \dfrac{X_L}{Z_{LR}}$ and $\cos \phi_1 = \dfrac{R}{Z_{LR}}$

$\tan \phi = \dfrac{I_{LR} \sin \phi_1 \sim I_C}{I_{LR} \cos \phi_1}$ and $\cos \phi = \dfrac{I_{LR} \cos \phi_1}{I}$

9 For any parallel a.c. circuit:
True or active power, $P = VI \cos \phi$ watts (W)
or $P = I_R^2 R$ watts
Apparent power, $S = VI$ voltamperes (VA)
Reactive power, $Q = VI \sin \phi$ reactive voltamperes (VAr)

Power factor $= \dfrac{\text{true power}}{\text{apparent power}} = \dfrac{P}{S} = \cos \phi$

(These formulae are the same as for series a.c. circuits.)
(see *Problems 1 to 7*)

10 (i) **Resonance** occurs in the two branch circuit containing capacitance C in parallel with inductance L and resistance R in series (see *Fig 2(a)*) when the quadrature (i.e. vertical) component of current I_{LR} is equal to I_C. At this condition the supply current I is in-phase with the supply voltage V.

(ii) When the quadrature component of I_{LR} is equal to I_C then:

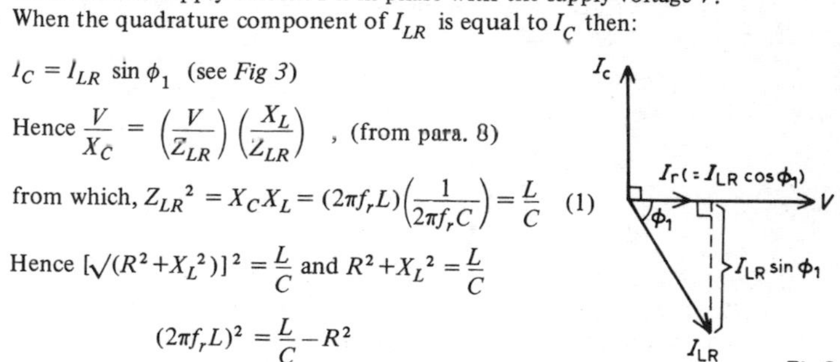

$I_C = I_{LR} \sin \phi_1$ (see *Fig 3*)

Hence $\dfrac{V}{X_C} = \left(\dfrac{V}{Z_{LR}}\right)\left(\dfrac{X_L}{Z_{LR}}\right)$, (from para. 8)

from which, $Z_{LR}^2 = X_C X_L = (2\pi f_r L)\left(\dfrac{1}{2\pi f_r C}\right) = \dfrac{L}{C}$ (1)

Hence $[\sqrt{(R^2 + X_L^2)}]^2 = \dfrac{L}{C}$ and $R^2 + X_L^2 = \dfrac{L}{C}$

$$(2\pi f_r L)^2 = \frac{L}{C} - R^2$$

Fig 3

$$f_r = \frac{1}{2\pi L}\sqrt{\left(\frac{L}{C} - R^2\right)} = \frac{1}{2\pi}\sqrt{\left(\frac{L}{L^2 C} - \frac{R^2}{L^2}\right)}$$

i.e. parallel resonant frequency, $f_r = \dfrac{1}{2\pi}\sqrt{\left(\dfrac{1}{LC} - \dfrac{R^2}{L^2}\right)}$ Hz

(When R is negligible, then $f_r = \dfrac{1}{2\pi\sqrt{(LC)}}$, which is the same as for series resonance.)

(iii) Current at resonance,

$$I_r = I_{LR} \cos \phi_1 \quad \text{(from } \textit{Fig 3}\text{)}$$

$$= \left(\frac{V}{Z_{LR}}\right)\left(\frac{R}{Z_{LR}}\right) \quad \text{(from para 8)}$$

$$= \frac{VR}{Z_{LR}^{\ 2}}$$

However from equation (1), $Z_{LR}^{\ 2} = \frac{L}{C}$

$$\text{Hence } I_r = \frac{VR}{\frac{L}{C}} = \frac{VRC}{L} \qquad (2)$$

The current is at a minimum at resonance.

(iv) Since the current at resonance is in-phase with the voltage the impedance of the circuit acts as a resistance. This resistance is known as the **dynamic resistance, R_D** (or sometimes, the dynamic impedance).

From equation (2), impedance at resonance $= \frac{V}{I_r} = \frac{L}{RC}$

i.e. dynamic resistance, $\mathbf{R_D = \frac{L}{RC}}$ **ohms**

(v) The parallel resonant circuit is often described as a **rejector** circuit since it presents its maximum impedance at the resonant frequency and the resultant current is a minimum.

11 Currents higher than the supply current can circulate within the parallel branches of a parallel resonant circuit, the current leaving the capacitor and establishing the magnetic field of the inductor, this then collapsing and recharging the capacitor, and so on. The **Q-factor** of a parallel resonant circuit is the ratio of the current circulating in the parallel branches of the circuit to the supply current, i.e. the current magnification.

$$\text{Q-factor at resonance} = \text{current magnification} = \frac{\text{circulating current}}{\text{supply current}}$$

$$= \frac{I_C}{I_r} = \frac{I_{LR} \sin \phi_1}{I_r}$$

$$= \frac{I_{LR} \sin \phi_1}{I_{LR} \cos \phi_1} = \frac{\sin \phi_1}{\cos \phi_1} = \tan \phi_1 = \frac{X_L}{R}$$

i.e. **Q-factor at resonance** $= \frac{2\pi f_r L}{R}$ (which is the same as for a series circuit)

(see *Problems 8 to 10*).

Note that in a parallel circuit the Q-factor is a measure of current magnification, whereas in a series circuit it is a measure of voltage magnification.

12 At mains frequencies the Q-factor of a parallel circuit is usually low, typically less than 10, but in radio-frequency circuits the Q-factor can be very high.

13 For a particular power supplied, a high power factor reduces the current flowing in a supply system and therefore reduces the cost of cables, switch-gear, transformers and generators. Supply authorities use tariffs which encourage electricity consumers to operate at a reasonably high power factor.

Industrial loads such as a.c. motors are essentially inductive (R–L) and may have a low power factor. One method of improving (or correcting) the power factor of an inductive load is to connect a static capacitor C in parallel with the

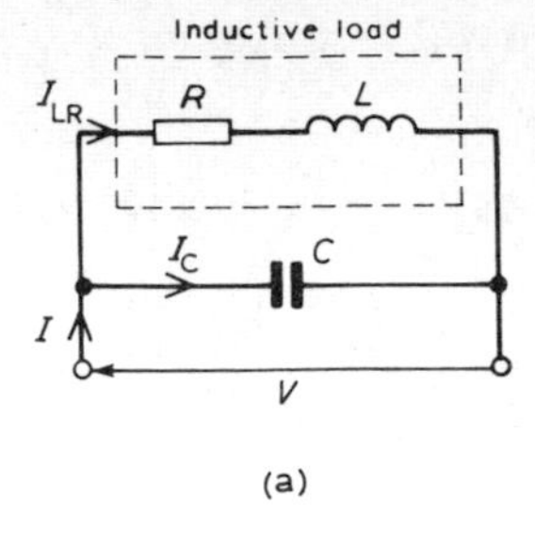

(a)

Fig 4

(b)

load (see *Fig 4(a)*). The supply current is reduced from I_{LR} to I, the phasor sum of I_{LR} and I_C, and the circuit power factor improves from $\cos \phi_1$ to $\cos \phi_2$ (see *Fig 4(b)*).
(See *Problems 11 to 14*).

B. WORKED PROBLEMS ON PARALLEL A.C. CIRCUITS

Problem 1 A 20 Ω resistor is connected in parallel with an inductance of 2.387 mH across a 60 V, 1 kHz supply. Calculate (a) the current in each branch: (b) the supply current; (c) the circuit phase angle; (d) the circuit impedance; and (e) the power consumed.

The circuit and phasor diagrams are as shown in *Fig 1(a)*.

(a) Current flowing in the resistor, $I_R = \frac{V}{R} = \frac{60}{20} = 3$ **A**

Current flowing in the inductance, $I_L = \frac{V}{X_L} = \frac{V}{2\pi f L}$

$$= \frac{60}{2\pi(1000)(2.387 \times 10^{-3})} = 4 \text{ A}$$

(b) From the phasor diagram, supply current, $I = \sqrt{(I_R^2 + I_L^2)} = \sqrt{(3^2 + 4^2)} = 5$ **A**

(c) Circuit phase angle $\phi = \arctan \frac{I_L}{I_R} = \arctan \frac{4}{3} = $ **53° 8′ lagging**

(d) Circuit impedance, $Z = \frac{V}{I} = \frac{60}{5} = 12$ **Ω**

(e) Power consumed $P = VI \cos \phi = (60)(5)(\cos 53° 8') = $ **180 W**
(Alternatively, power consumed $P = I_R^2 R = (3)^2(20) = $ **180 W**)

Problem 2 A 30 μF capacitor is connected in parallel with an 80 Ω resistor across a 240 V, 50 Hz supply. Calculate (a) the current in each branch; (b) the supply current; (c) the circuit phase angle; (d) the circuit impedance; (e) the power dissipated and (f) the apparent power.

The circuit and phasor diagrams are as shown in *Fig 1(b)*.

(a) Current in resistor $I_R = \frac{V}{R} = \frac{240}{80} = \mathbf{3\ A}$

Current in capacitor $I_C = \frac{V}{X_C} = \frac{V}{\frac{1}{2\pi fC}} = 2\pi fCV = 2\pi(50)(30 \times 10^{-6})(240)$

$= \mathbf{2.262\ A}$

(b) Supply current $I = \sqrt{(I_R^2 + I_C^2)} = \sqrt{(3^2 + 2.262^2)} = \mathbf{3.757\ A}$

(c) Circuit phase angle, $\alpha = \arctan \frac{I_C}{I_R} = \arctan \frac{2.262}{3} = \mathbf{37^\circ\ 1'\ leading}$

(d) Circuit impedance $Z = \frac{V}{I} = \frac{240}{3.757} = \mathbf{63.88\ \Omega}$

(e) True or active power dissipated $P = VI \cos \alpha = (240)(3.757) \cos 37^\circ\ 1'$

$= \mathbf{720\ W}$

(Alternatively, true power $P = I_R^2 R = (3)^2(80) = \mathbf{720\ W}$)

(f) Apparent power, $S = VI = (240)(3.757) = \mathbf{901.7\ VA}$

Problem 3 A capacitor C is connected in parallel with a resistor R across a 120 V, 200 Hz supply. The supply current is 2 A at a power factor of 0.6 leading. Determine the values of C and R.

The circuit diagram is shown in *Fig 5(a)*.

Power factor $= \cos \phi = 0.6$ leading. Hence $\phi = \arccos 0.6 = 53^\circ\ 8'$ leading.

From the phasor diagram shown in *Fig 5(b)*, $I_R = I \cos 53^\circ\ 8' = (2)(0.6) = 1.2$ A

and $I_C = I \sin 53^\circ\ 8' = (2)(0.8) = 1.6$ A

(Alternatively, I_R and I_C can be measured from a scaled phasor diagram.)

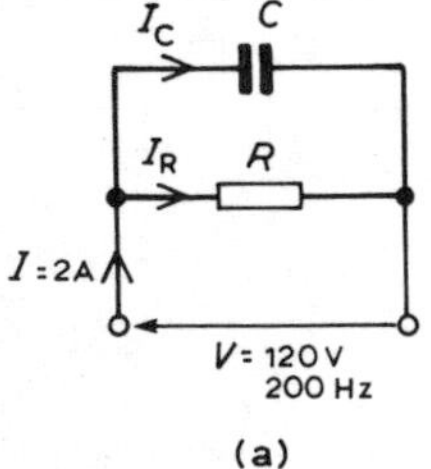

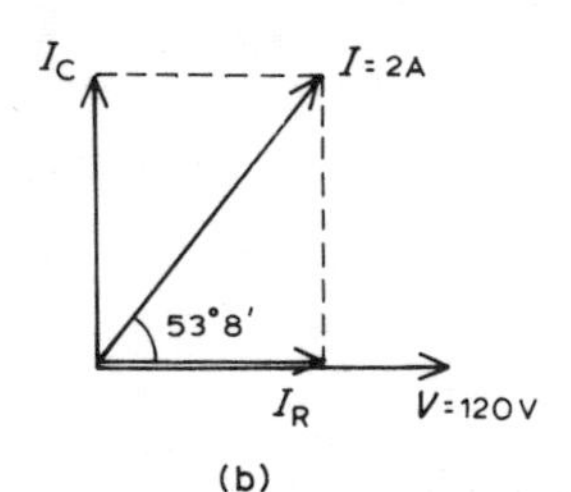

Fig 5

From the circuit diagram, $I_R = \frac{V}{R}$ from which $R = \frac{V}{I_R} = \frac{120}{1.2} = \mathbf{100\ \Omega}$

and $I_C = \frac{V}{X_C} = 2\pi fCV$, from which $C = \frac{I_C}{2\pi fV}$

$= \frac{1.6}{2\pi(200)(120)}$

$= \mathbf{10.61\ \mu F}$

Problem 4 A pure inductance of 120 mH is connected in parallel with a 25 μF capacitor and the network is connected to a 100 V, 50 Hz supply. Determine (a) the branch currents; (b) the supply current and its phase angle; (c) the circuit impedance; and (d) the power consumed.

The circuit and phasor diagrams are as shown in *Fig 1(c)*.

(a) Inductive reactance $X_L = 2\pi fL = 2\pi(50)(120 \times 10^{-3}) = 37.70\ \Omega$

Capacitive reactance $X_C = \dfrac{1}{2\pi fC} = \dfrac{1}{2\pi(50)(25 \times 10^{-6})} = 127.3\ \Omega$

Current flowing in inductance $I_L = \dfrac{V}{X_L} = \dfrac{100}{37.70} = \mathbf{2.653\ A}$

Current flowing in capacitor $I_C = \dfrac{V}{X_C} = \dfrac{100}{127.3} = \mathbf{0.786\ A}$

(b) I_L and I_C are anti-phase. Hence supply current $I = I_L - I_C = 2.653 - 0.786 = \mathbf{1.867\ A}$

The current I lags the supply voltage V by 90° (see *Fig 1(c)(i)*.

(c) Circuit impedance $Z = \dfrac{V}{I} = \dfrac{100}{1.867} = \mathbf{53.56\ \Omega}$

(d) Power consumed $P = VI \cos\phi = (100)(1.867)(\cos 90°) = \mathbf{0\ W}$

Problem 5 Repeat Problem 4 for the condition when the frequency is changed to 150 Hz

(a) Inductive reactance, $X_L = 2\pi(150)(120 \times 10^{-3}) = 113.1\ \Omega$

Capacitive reactance, $X_C = \dfrac{1}{2\pi(150)(25 \times 10^{-6})} = 42.44\ \Omega$

Current flowing in inductance, $I_L = \dfrac{V}{X_L} = \dfrac{100}{113.1} = \mathbf{0.884\ A}$

Current flowing in capacitor, $I_C = \dfrac{V}{X_C} = \dfrac{100}{42.44} = \mathbf{2.356\ A}$

(b) Supply current, $I = I_C - I_L = 2.356 - 0.884 = $ **1.472 A leading V by 90°** (see *Fig 1(c)(ii)*.

(c) Circuit impedance, $Z = \dfrac{V}{I} = \dfrac{100}{1.472} = \mathbf{67.93\ \Omega}$

(d) Power consumed, $P = VI \cos\phi = \mathbf{0\ W}$ (since $\phi = 90°$)

From Problems 4 and 5:

(i) When $X_L < X_C$ then $I_L > I_C$ and I lags V by 90°.
(ii) When $X_L > X_C$ then $I_L < I_C$ and I leads V by 90°.
(iii) In a parallel circuit containing no resistance the power consumed is zero.

Problem 6 A coil of inductance 159.2 mH and resistance 40 Ω is connected in parallel with a 30 μF capacitor across a 240 V, 50 Hz supply. Calculate (a) the current in the coil and its phase angle, (b) the current in the capacitor and its phase angle, (c) the supply current and its phase angle, (d) the circuit impedance, (e) the power consumed, (f) the apparent power and (g) the reactive power. Draw the phasor diagram.

The circuit diagram is shown in *Fig 6(a)*.

(a) For the coil, inductive reactance $X_L = 2\pi fL = 2\pi(50)(159.2 \times 10^{-3}) = 50\ \Omega$
Impedance $Z_1 = \sqrt{(R^2 + X_L^2)} = \sqrt{(40^2 + 50^2)} = 64.03\ \Omega$

Current in coil, $I_{LR} = \dfrac{V}{Z_1} = \dfrac{240}{64.03} = \mathbf{3.748\ A}$

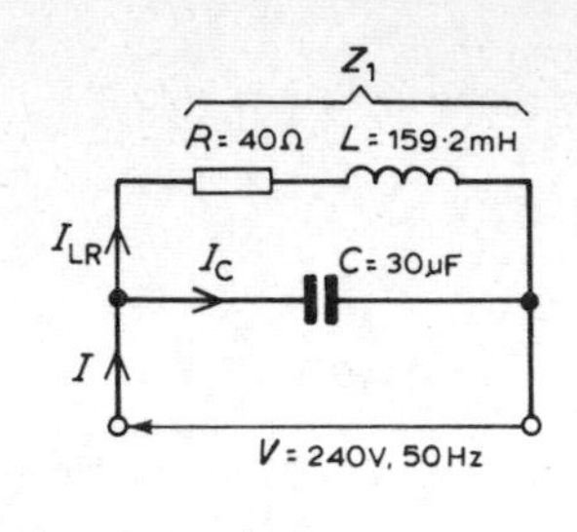

(a) Fig 6 (b)

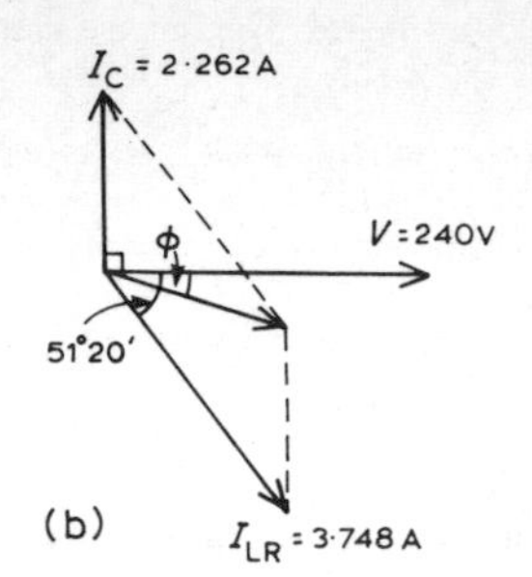

Branch phase angle $\phi_1 = \arctan \dfrac{X_L}{R} = \arctan \dfrac{50}{40} = \arctan 1.25$
$= \mathbf{51^\circ\ 20'}$ **lagging** (see phasor diagram in *Fig 6(b)*)

(b) Capacitive reactance, $X_C = \dfrac{1}{2\pi fC} = \dfrac{1}{2\pi(50)(30 \times 10^{-6})} = 106.1\ \Omega$

Current in capacitor, $I_C = \dfrac{V}{X_C} = \dfrac{240}{106.1} = \mathbf{2.262\ A}$ **leading the supply voltage by 90°** (see phasor diagram of *Fig 6(b)*).

(c) The supply current I is the phasor sum of I_{LR} and I_C. This may be obtained by drawing the phasor diagram to scale and measuring the current I and its phase angle relative to V. (Current I will always be the diagonal of the parallelogram formed as in *Fig 6(b)*). Alternatively the currents I_{LR} and I_C may be resolved into their horizontal (or 'in-phase') and vertical (or 'quadrature') components. The horizontal component of I_{LR} is

$I_{LR} \cos(51^\circ\ 20') = 3.748 \cos 51^\circ\ 20' = 2.342$ A

The horizontal component of I_C is $I_C \cos 90^\circ = 0$.

Thus the total horizontal component, $I_H = \mathbf{2.342\ A}$

The vertical component of $I_{LR} = -I_{LR} \sin(51^\circ\ 20') = -3.748 \sin 51^\circ\ 20'$
$= -2.926$ A

The vertical component of $I_C = I_C \sin 90^\circ = 2.262 \sin 90^\circ = 2.262$ A
Thus the total vertical component, $I_V = -2.926 + 2.262 = \mathbf{-0.664\ A}$
I_H and I_V are shown in *Fig 7*, from which,

$I = \sqrt{[(2.342)^2 + (-0.664)^2]} = 2.434$ A

Angle $\phi = \arctan \dfrac{0.664}{2.342} = 15^\circ\ 50'$ lagging.

Hence the supply current I = 2.434 A lagging V by 15° 50′.

Fig 7

(d) Circuit impedance $Z = \dfrac{V}{I} = \dfrac{240}{2.434} = \mathbf{98.60\ \Omega}$

(e) Power consumed $P = VI \cos\phi = (240)(2.434) \cos 15^\circ\ 50' = \mathbf{562\ W}$
(Alternatively $P = I_R{}^2 R = I_{LR}{}^2 R$ (in this case) $= (3.748)^2(40) = 562$ W)

(f) Apparent power $S = VI = (240)(2.434) = \mathbf{584.2\ V\,A}$

(g) Reactive power $Q = VI \sin\phi = (240)(2.434)(\sin 15^\circ\ 50') = \mathbf{159.4\ var}$

Problem 7 A coil of inductance 0.12 H and resistance 3 kΩ is connected in parallel with a 0.02 μF capacitor and is supplied at 40 V at a frequency of 5 kHz. Determine (a) the current in the coil, and (b) the current in the capacitor. (c) Draw to scale the phasor diagram and measure the supply current and its phase angle. Calculate (d) the circuit impedance and (e) the power consumed.

The circuit diagram is shown in *Fig 8(a)*.

(a) Inductive reactance, $X_L = 2\pi fL = 2\pi(5000)(0.12) = 3770\ \Omega$

Impedance of coil, $Z_1 = \sqrt{(R^2 + X_L^2)} = \sqrt{[(3000)^2 + (3770)^2]} = 4818\ \Omega$

Current in coil, $I_{LR} = \dfrac{V}{Z_1} = \dfrac{40}{4818} = \mathbf{8.30\ mA}$

Branch phase angle $\phi = \arctan \dfrac{X_L}{R} = \arctan \dfrac{3770}{3000} = \mathbf{51.5^\circ}$ **lagging.**

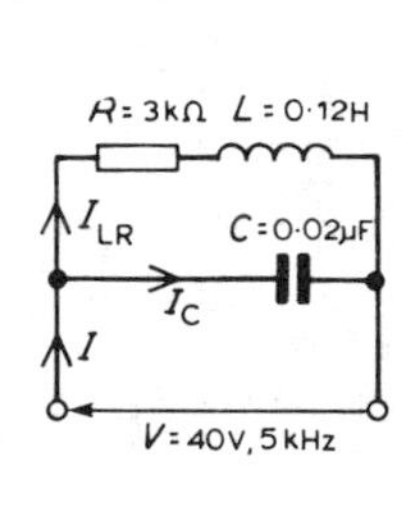

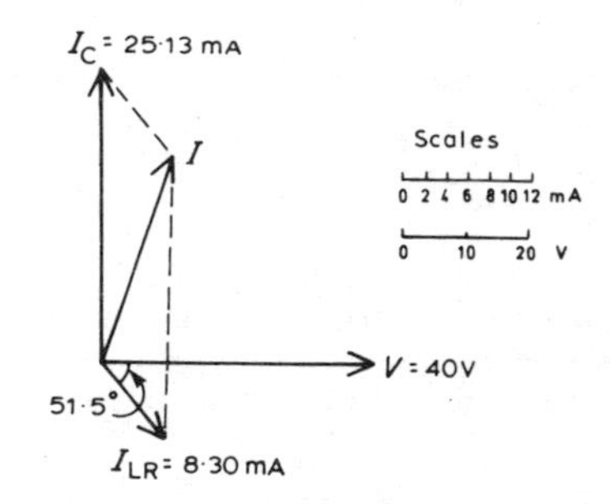

Fig 8

(b) Capacitive reactance, $X_C = \dfrac{1}{2\pi fC} = \dfrac{1}{2\pi(5000)(0.02 \times 10^{-6})} = 1592\ \Omega$

Capacitor current, $I_C = \dfrac{V}{X_C} = \dfrac{40}{1592} = \mathbf{25.13\ mA}$ **leading V by 90°.**

(c) Currents I_{LR} and I_C are shown in the phasor diagram of *Fig 8(b)*. The parallelogram is completed as shown and the supply current is given by the diagonal of the parallelogram. The current I is measured as **19.3 mA** leading voltage V by **74.5°**

(By calculation, $I = \sqrt{[(I_{LR} \cos 51.5^\circ)^2 + (I_C - I_{LR} \sin 51.5^\circ)^2]} = 19.34$ A

and $\phi = \arctan \left(\dfrac{I_C - I_{LR} \sin 51.5^\circ}{I_{LR} \cos 51.5^\circ} \right) = 74.5^\circ$

(d) Circuit impedance, $Z = \dfrac{V}{I} = \dfrac{40}{19.3 \times 10^{-3}} = \mathbf{2.073\ k\Omega}$

(e) Power consumed, $P = VI \cos \phi = (40)(19.3 \times 10^{-3})(\cos 74.5^\circ) = \mathbf{206.3\ mW}$

(Alternatively, $P = I_R^2 R = I_{LR}^2 R = (8.30 \times 10^{-3})^2 (3000) = 206.7$ mW)

Problem 8 A pure inductance of 150 mH is connected in parallel with a 40 μF capacitor across a 50 V, variable frequency supply. Determine (a) the resonant frequency of the circuit and (b) the current circulating in the capacitor and inductance at resonance.

The circuit diagram is shown in *Fig 9*.

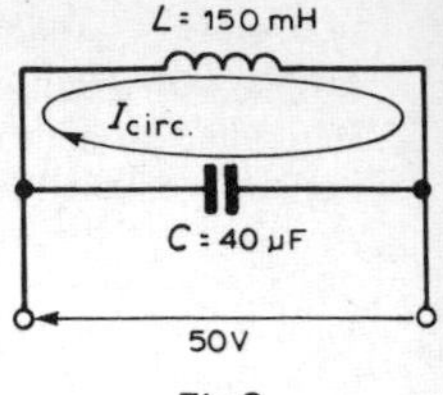

Fig 9

(a) Parallel resonant frequency, $f_r = \frac{1}{2\pi}\sqrt{\left(\frac{1}{LC} - \frac{R^2}{L^2}\right)}$

However, resistance $R = 0$. Hence,

$$f_r = \frac{1}{2\pi}\sqrt{\left(\frac{1}{LC}\right)} = \frac{1}{2\pi}\sqrt{\left(\frac{1}{(150 \times 10^{-3})(40 \times 10^{-6})}\right)}$$

$$= \frac{1}{2\pi}\sqrt{\left(\frac{10^7}{(15)(4)}\right)} = \frac{10^3}{2\pi}\sqrt{\left(\frac{1}{6}\right)} = \mathbf{64.97\ Hz.}$$

(b) Current circulating in L and C at resonance,

$$I_{CIRC} = \frac{V}{X_C} = \frac{V}{\frac{1}{2\pi f_r C}} = 2\pi f_r CV$$

Hence $I_{CIRC} = 2\pi(64.97)(40 \times 10^{-6})(50) = \mathbf{0.816\ A}$

(Alternatively, $I_{CIRC} = \frac{V}{X_L} = \frac{V}{2\pi f_r L} = \frac{50}{2\pi(64.97)(0.15)} = \mathbf{0.817\ A}$)

Problem 9 A coil of inductance 0.20 H and resistance 60 Ω is connected in parallel with a 20 μF capacitor across a 20 V, variable frequency supply. Calculate (a) the resonant frequency; (b) the dynamic resistance; (c) the current at resonance and (d) the circuit Q-factor at resonance.

(a) Parallel resonant frequency,

$$f_r = \frac{1}{2\pi}\sqrt{\left(\frac{1}{LC} - \frac{R^2}{L^2}\right)}$$
$$= \frac{1}{2\pi}\sqrt{\left(\frac{1}{(0.20)(20 \times 10^{-6})} - \frac{(60)^2}{(0.2)^2}\right)}$$
$$= \frac{1}{2\pi}\sqrt{(250\,000 - 90\,000)} = \frac{1}{2\pi}\sqrt{(160\,000)}$$
$$= \frac{1}{2\pi}(400)$$
$$= \mathbf{63.66\ Hz}$$

(b) Dynamic resistance, $R_D = \frac{L}{RC} = \frac{0.2}{(60)(20 \times 10^{-6})} = \mathbf{166.7\ \Omega}$

(c) Current at resonance, $I_r = \frac{V}{R_D} = \frac{20}{166.7} = \mathbf{0.12\ A}$

(d) Circuit Q-factor at resonance $= \frac{2\pi f_r L}{R} = \frac{2\pi(63.66)(0.2)}{60} = \mathbf{1.33}$

Alternatively, Q-factor at resonance = current magnification (for a parallel circuit) $= I_c/I_r$

$$I_c = \frac{V}{X_c} = \frac{V}{\frac{1}{2\pi f_r C}} = 2\pi f_r CV = 2\pi(63.66)(20 \times 10^{-6})(20) = 0.16\ \text{A}$$

Hence Q-factor $= \frac{I_c}{I_r} = \frac{0.16}{0.12} = \mathbf{1.33}$, as obtained above.

Problem 10 A coil of inductance 100 mH and resistance 800 Ω is connected in parallel with a variable capacitor across a 12 V, 5 kHz supply. Determine for the condition when the supply current is a minimum: (a) the capacitance of the capacitor, (b) the dynamic resistance, (c) the supply current, and (d) the Q-factor.

(a) The supply current is a minimum when the parallel circuit is at resonance.

$$\text{Resonant frequency, } f_r = \frac{1}{2\pi}\sqrt{\left(\frac{1}{LC} - \frac{R^2}{L^2}\right)}$$

Transposing for C gives: $$(2\pi f_r)^2 = \frac{1}{LC} - \frac{R^2}{L^2}$$

$$(2\pi f_r)^2 + \frac{R^2}{L^2} = \frac{1}{LC}$$

$$C = \frac{1}{L\left\{(2\pi f_r)^2 + \frac{R^2}{L^2}\right\}}$$

When $L = 100$ mH, $R = 800$ Ω and $f_r = 5000$ Hz,

$$C = \frac{1}{100 \times 10^{-3}\left\{(2\pi 5000)^2 + \frac{800^2}{(100 \times 10^{-3})^2}\right\}}$$

$$= \frac{1}{0.1[\pi^2 10^8 + (0.64)10^8]} \text{ F}$$

$$= \frac{10^6}{0.1(10.51 \times 10^8)} \ \mu\text{F}$$

$= \mathbf{0.009\ 515\ \mu F \text{ or } 9.515\ nF.}$

(b) Dynamic resistance, $R_D = \dfrac{L}{CR} = \dfrac{100 \times 10^{-3}}{(9.515 \times 10^{-9})(800)} = \mathbf{13.14\ k\Omega}$

(c) Supply current at resonance, $I_r = \dfrac{V}{R_D} = \dfrac{12}{13.14 \times 10^3} = \mathbf{0.913\ mA.}$

(d) Q-factor at resonance $= \dfrac{2\pi f_r L}{R} = \dfrac{2\pi(5000)(100 \times 10^{-3})}{800} = \mathbf{3.93}$

Alternatively, Q-factor at resonance $= \dfrac{I_c}{I_r} = \dfrac{V/X_c}{I_r} = \dfrac{2\pi f_r CV}{I_r}$

$$= \frac{2\pi(5000)(9.515 \times 10^{-9})(12)}{0.913 \times 10^{-3}} = \mathbf{3.93}$$

Problem 11 A single-phase motor takes 50 A at a power factor of 0.6 lagging from a 240 V, 50 Hz supply. Determine (a) the current taken by a capacitor connected in parallel with the motor to correct the power factor to unity, and (b) the value of the supply current after power factor correction.

The circuit diagram is shown in *Fig 10(a)*.

(a) A power factor of 0.6 lagging means that $\cos\phi = 0.6$
i.e. $\phi = \arccos 0.6 = 53° 8'$.
Hence I_M lags V by 53° 8′ as shown in *Fig 10(b)*.
If the power factor is to be improved to unity then the phase difference

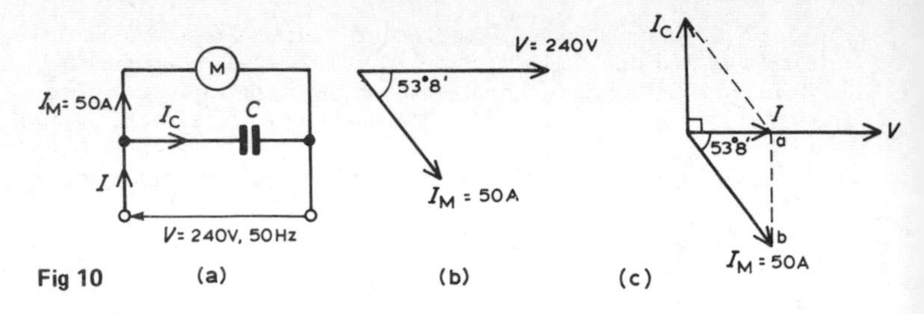

Fig 10 (a) (b) (c)

between supply current I and voltage V is 0°, i.e. I is in phase with V as shown in *Fig 10(c)*. For this to be so, I_C must equal the length ab, such that the phasor sum of I_M and I_C is I.

ab = $I_M \sin 53° \, 8' = 50(0.8) = 40$ A.

Hence the capacitor current I_c must be 40 A for the power factor to be unity.

(b) Supply current $I = I_M \cos 53° \, 8' = 50(0.6) =$ **30 A.**

Problem 12 A 400 V alternator is supplying a load of 42 kW at a power factor of 0.7 lagging. Calculate (a) the kVA loading and (b) the current taken from the alternator. (c) If the power factor is now raised to unity find the new kVA loading

(a) Power = $VI \cos \phi = (VI)$ (power factor)

Hence $VI = \dfrac{\text{power}}{\text{p.f.}} = \dfrac{42 \times 10^3}{0.7} =$ **60 kVA**

(b) $VI = 60\,000$ V A. Hence $I = \dfrac{60\,000}{V} = \dfrac{60\,000}{400} =$ **150 A**

(c) The kVA loading remains at **60 kVA** irrespective of changes in power factor.

Problem 13 A motor has an output of 4.8 kW, an efficiency of 80% and a power factor of 0.625 lagging when operated from a 240 V, 50 Hz supply. It is required to improve the power factor to 0.95 lagging by connecting a capacitor in parallel with the motor. Determine (a) the current taken by the motor; (b) the supply current after power factor correction; (c) the current taken by the capacitor; (d) the capacitance of the capacitor, and (e) the kvar rating of the capacitor.

(a) Efficiency = $\dfrac{\text{power output}}{\text{power input}}$. Hence $\dfrac{80}{100} = \dfrac{4800}{\text{power input}}$

Power input = $\dfrac{4800}{0.8} = 6000$ W.

Hence, $6000 = VI_M \cos \phi = (240)(I_M)(0.625)$, since $\cos \phi =$ p.f. = 0.625

Thus current taken by the motor, $I_M = \dfrac{6000}{(240)(0.625)} =$ **40 A**

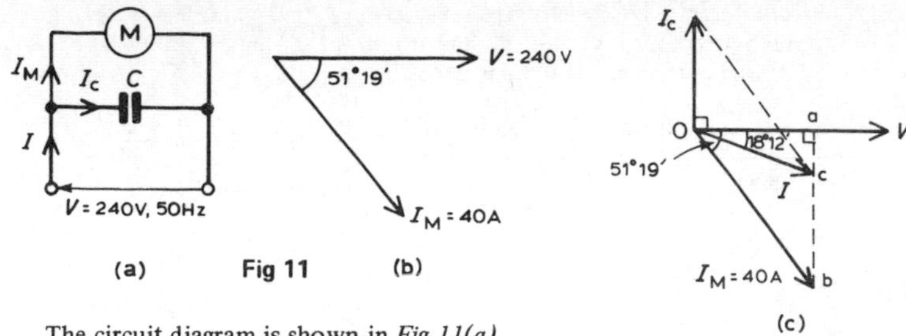

Fig 11

The circuit diagram is shown in *Fig 11(a)*.
The phase angle between I_M and V is given by:
$\phi = \arccos 0.625 = 51° \ 19'$ hence the phasor diagram is as shown in *Fig 11(b)*.

(b) When a capacitor C is connected in parallel with the motor a current I_C flows which leads V by 90°. The phasor sum of I_M and I_C gives the supply current I, and has to be such as to change the circuit power factor to 0.95 lagging, i.e. a phase angle of arccos 0.95 or 18° 12′ lagging, as shown in *Fig 11(c)*.
The horizontal component of I_M (shown as oa) $= I_M \cos 51° \ 19'$
$= 40 \cos 51° \ 19' = 25$ A
The horizontal component of I (also given by oa) $= I \cos 18° \ 12' = 0.95\ I$
Equating the horizontal components gives: $25 = 0.95\ I$

Hence the supply current after p.f. correction, $I = \dfrac{25}{0.95} = $ **26.32 A**

(c) The vertical component of I_M (shown as ab) $= I_M \sin 51° \ 19'$
$= 40 \sin 51° \ 19' = 31.22$ A
The vertical component of I (shown as ac) $= I \sin 18° \ 12'$
$= 26.32 \sin 18° \ 12' = 8.22$ A
The magnitude of the capacitor current I_C (shown as bc) is given by ab–ac, i.e. 31.22–8.22 = **23 A**

(d) Current $I_C = \dfrac{V}{X_C} = \dfrac{V}{\dfrac{1}{2\pi fC}} = 2\pi fCV,$

from which $C = \dfrac{I_C}{2\pi fV} = \dfrac{23}{2\pi(50)(240)}$ F = **305 μF**

(e) kvar rating of the capacitor $= \dfrac{VI_C}{1000} = \dfrac{(240)(23)}{1000} = $ **5.52 kvar**

In this problem the supply current has been reduced from 40 A to 26.32 A without altering the current or power taken by the motor. This means that the size of generating plant and the cross-sectional area of conductors supplying both the factory and the motor can be less—with an obvious saving in cost.

Problem 14 A 250 V, 50 Hz single-phase supply feeds the following loads (i) incandescent lamps taking a current of 10 A at unity power factor; (ii) fluorescent lamps taking 8 A at a power factor of 0.7 lagging; (iii) a 3 kVA motor operating at full load and at a power factor of 0.8 lagging and (iv) a static capacitor. Determine, for the lamps and motor, (a) the total current; (b) the overall power factor and (c) the total power, (d) Find the value of the static capacitor to improve the overall power factor to 0.975 lagging.

A phasor diagram is constructed as shown in *Fig 12(a)*, where 8 A is lagging voltage V by arccos 0.7, i.e. 45.57°, and the motor current is 3000/250, i.e. 12 A lagging V by arccos 0.8, i.e. 36.87°.

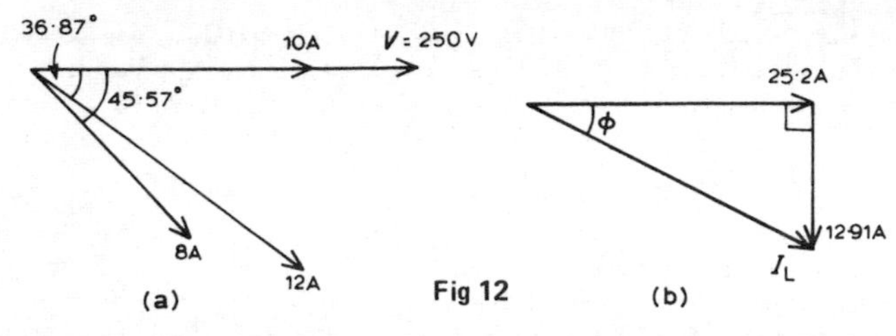

Fig 12

(a) The horizontal component of the currents $= 10\cos 0° + 12\cos 36.87° + 8\cos 45.57°$
$= 10+9.6+5.6 = 25.2$ A.

The vertical component of the currents $= 10\sin 0°+12\sin 36.87° + 8\sin 45.57°$
$= 0+7.2+5.713 = 12.91$ A

From *Fig 12(b)*, total current, $I_L = \sqrt{[(25.2)^2+(12.91)^2]} =$ **28.31 A**

at a phase angle of $\phi = \arctan\left(\frac{12.91}{25.2}\right)$, i.e. 27.13° lagging.

(b) Power factor $= \cos\phi = \cos 27.13° =$ **0.890 lagging.**

(c) Total power, $P = VI_L\cos\phi = (250)(28.31)(0.890) =$ **6.3 kW**

(d) To improve the power factor, a capacitor is connected in parallel with the loads. The capacitor takes a current I_C such that the supply current falls from 28.31 A to I, lagging V by arccos 0.975, i.e. 12.84°. The phasor diagram is shown in *Fig 13*.

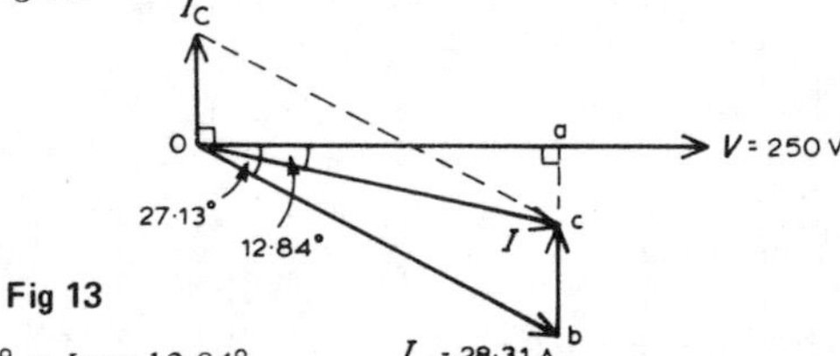

Fig 13

oa $= 28.31\cos 27.13° = I\cos 12.84°$

Hence $I = \frac{28.31\cos 27.13°}{\cos 12.84°} = 25.84$ A

Current $I_C = \text{bc} = (\text{ab}-\text{ac}) = 28.31\sin 27.13° - 25.84\sin 12.84°$
$= 12.91-5.742 = 7.168$ A

$$I_C = \frac{V}{X_C} = \frac{V}{\frac{1}{2\pi fC}} = 2\pi fCV$$

Hence capacitance $C = \frac{I_C}{2\pi fV} = \frac{7.168}{2\pi(50)(250)}$ **F = 91.27 πF**

Thus to improve the power factor from 0.890 to 0.975 lagging a 91.27 μF capacitor is connected in parallel with the loads.

C. FURTHER PROBLEMS ON PARALLEL A.C. CIRCUITS

SHORT ANSWER PROBLEMS

1. Draw the phasor diagram for a two-branch parallel circuit containing capacitance C in one branch and resistance R in the other, connected across a supply voltage V.
2. Draw the phasor diagram for a two-branch parallel circuit containing inductance L and resistance R in series in one branch, and capacitance C in the other, connected across a supply voltage V.
3. Draw the phasor diagram for a two-branch parallel circuit containing inductance L in one branch and capacitance C in the other for the condition in which inductive reactance is greater than capacitive reactance.
4. State two methods of determining the phasor sum of two currents.
5. State two formulae which may be used to calculate power in a parallel a.c. circuit.
6. State the condition for resonance for a two-branch circuit containing capacitance C in parallel with a coil of inductance L and resistance R.
7. Develop a formula for parallel resonance in terms of resistance R, inductance L and capacitance C.
8. What does the Q-factor of a parallel circuit mean?
9. Develop a formula for the current at resonance in terms of resistance R, inductance L, capacitance C and supply voltage V.
10. What is dynamic resistance? State a formula for dynamic resistance.
11. Explain a simple method of improving the power factor of an inductive circuit.
12. Why is it advantageous to improve power factor?

MULTI-CHOICE PROBLEMS (answers on page 191)

A 2-branch parallel circuit, containing a 10 Ω resistance in one branch and a 100 μF capacitor in the other, has a 120 V, $2/3\pi$ kHz supply connected across it. Determine the quantities stated in *Problems 1 to 8*, selecting the correct answer from the following list.

(a) 24 A; (b) 6 Ω; (c) 7.5 kΩ; (d) 12 A; (e) arctan 3/4 leading; (f) 0.8 leading; (g) 7.5 Ω; (h) arctan 4/3 leading; (i) 16 A; (j) arctan 5/3 lagging; (k) 1.44 kW; (l) 0.6 leading; (m) 22.5 Ω; (n) 2.4 kW; (o) arctan 4/3 lagging; (p) 0.6 lagging; (q) 0.8 lagging; (r) 1.92 kW; (s) 20 A.

1. The current flowing in the resistance.
2. The capacitive reactance of the capacitor.
3. The current flowing in the capacitor.
4. The supply current.
5. The supply phase angle.
6. The circuit impedance.
7. The power consumed by the circuit.
8. The power factor of the circuit.
9. A 2 branch parallel circuit consists of a 15 mH inductance in one branch and a 50 μF capacitor in the other across a 120 V, $1/\pi$ kHz supply. The supply current is:

   (a) 8 A leading by $\pi/2$ rads (b) 16 A lagging by 90°
   (c) 8 A lagging by 90° (d) 16 A leading by $\pi/2$ rads.

10 The following statements, taken correct to 2 significant figures, refer to the circuit shown in *Fig 14*. Which are false?

(a) The impedance of the R–L branch is 5 Ω.

(b) $I_{LR} = 50$ A; (c) $I_C = 20$ A; (d) $L = 0.80$ H;

(e) $C = 16\ \mu$F; (f) The 'in-phase' component of the supply current is 30 A

(g) The 'quadrature' component of the supply current is 40 A (h) $I = 36$ A. (i) Circuit phase angle = 33° 41′ leading.

(j) Circuit impedance = 6.9 Ω;

(k) Circuit power factor = 0.83 lagging.

(l) Power consumed = 9.0 kW

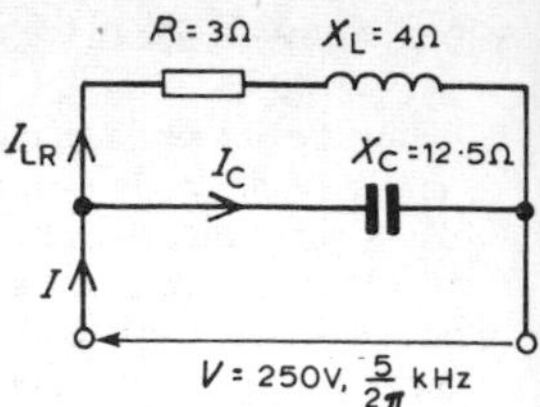

Fig 14

11 Which of the following statements is false?

(a) The supply current is a minimum at resonance in a parallel circuit.

(b) The Q-factor at resonance in a parallel circuit is the voltage magnification.

(c) Improving power factor reduces the current flowing through a system.

(d) The circuit impedance is a maximum at resonance in a parallel circuit.

12 An LR–C parallel circuit (similar to *Fig 12*) has the following component values: $R = 10$ Ω, $L = 10$ mH, $C = 10\ \mu$F, $V = 100$ V. Which of the following statements is false?

(a) The resonant frequency f_r is $1.5/\pi$ kHz.

(b) The current at resonance is 1 A.

(c) The dynamic resistance is 100 Ω.

(d) The circuit Q-factor at resonance is 30.

CONVENTIONAL PROBLEMS

1 A 30 Ω resistor is connected in parallel with a pure inductance of 3 mH across a 110 V, 2 kHz supply. Calculate (a) the current in each branch; (b) the circuit current; (c) the circuit phase angle; (d) the circuit impedance; (e) the power consumed, and (f) the circuit power factor.

[(a) $I_R = 3.67$ A, $I_L = 2.92$ A (b) 4.69 A (c) 38° 30′ lagging (d) 23.45 Ω (e) 404 W (f) 0.783 lagging]

2 A 40 Ω resistance is connected in parallel with a coil of inductance L and negligible resistance across a 200 V, 50 Hz supply and the supply current is found to be 8 A. Draw a phasor diagram to scale and determine the inductance of the coil. [102 mH]

3 A 1500 nF capacitor is connected in parallel with a 16 Ω resistor across a 10 V, 10 kHz supply. Calculate (a) the current in each branch; (b) the supply current; (c) the circuit phase angle; (d) the circuit impedance; (e) the power consumed; (f) the apparent power; and (g) the circuit power factor. Draw the phasor diagram.

[(a) $I_R = 0.625$ A, $I_C = 0.942$ A; (b) 1.13 A; (c) 56° 28′ leading; (d) 8.85 Ω; (e) 6.25 W; (f) 11.3 VA; (g) 0.55 leading.]

4 A capacitor C is connected in parallel with a resistance R across a 60 V, 100 Hz supply. The supply current is 0.6 A at a power factor of 0.8 leading. Calculate the value of R and C. [$R = 125$ Ω; $C = 9.55\ \mu$F]

5 An inductance of 80 mH is connected in parallel with a capacitance of 10 μF across a 60 V, 100 Hz supply. Determine (a) the branch currents; (b) the supply

current; (c) the circuit phase angle; (d) the circuit impedance and (e) the power consumed.

[(a) $I_C = 0.377$ A, $I_L = 1.194$ A; (b) 0.817 A; (c) 90° lagging; (d) 73.44 Ω; (e) 0 W.]

6 Repeat problem 5 for a supply frequency of 200 Hz.

[(a) $I_C = 0.754$ A, $I_L = 0.597$ A; (b) 0.157 A; (c) 90° leading; (d) 382.2 Ω; (e) 0 W.]

7 A coil of resistance 60 Ω and inductance 318.4 mH is connected in parallel with a 15 μF capacitor across a 200 V, 50 Hz supply. Calculate (a) the current in the coil; (b) the current in the capacitor; (c) the supply current and its phase angle; (d) the circuit impedance; (e) the power consumed; (f) the apparent power and (g) the reactive power. Draw the phasor diagram.

[(a) 1.715 A; (b) 0.943 A; (c) 1.028 A at 30° 53′ lagging; (d) 194.6 Ω; (e) 176.4 W; (f) 205.6 VA; (g) 105.5 var.]

8 A 2/5 nF capacitor is connected in parallel with a coil of resistance 2 kΩ and inductance 0.20 H across a 100 V, 4 kHz supply. Determine (a) the current in the coil; (b) the current in the capacitor; (c) the supply current and its phase angle (by drawing a phasor diagram to scale, and also by calculation); (d) the circuit impedance; and (e) the power consumed.

[(a) 18.48 mA; (b) 62.83 mA; (c) 46.17 mA at 81° 29′ leading; (d) 2.166 k Ω; (e) 0.684 W.]

9 A 0.15 μF capacitor and a pure inductance of 0.01 H are connected in parallel across a 10 V, variable frequency supply. Determine (a) the resonant frequency of the circuit, and (b) the current circulating in the capacitor and inductance.

[(a) 4.11 kHz; (b) 38.74 mA]

10 A 30 μF capacitor is connected in parallel with a coil of inductance 50 mH and unknown resistance *R* across a 120 V, 50 Hz supply. If the circuit has an overall power factor of 1 find (a) the value of *R*; (b) the current in the coil and (c) the supply current. [(a) 37.7 Ω; (b) 2.94 A; (c) 2.714 A.]

11 A coil of resistance 25 Ω and inductance 150 mH is connected in parallel with a 10 μF capacitor across a 60 V, variable frequency supply. Calculate (a) the resonant frequency; (b) the dynamic resistance; (c) the current at resonance and (d) the Q-factor at resonance. [(a) 127.2 Hz (b) 600 Ω (c) 0.10 A (d) 4.80]

12 A coil having resistance *R* and inductance 80 mH is connected in parallel with a 5 nF capacitor across a 25 V, 3 kHz supply. Determine for the condition when the current is a minimum, (a) the resistance *R* of the coil; (b) the dynamic resistance; (c) the supply current; and (d) the Q-factor.

[(a) 3.705 kΩ; (b) 4.318 kΩ; (c) 5.79 mA; (d) 0.40.]

13 A 415 V alternator is supplying a load of 55 kW at a power factor of 0.65 lagging. Calculate (a) the kVA loading and (b) the current taken from the alternator. (c) If the power factor is now raised to unity find the new kVA loading.

[(a) 84.6 kVA; (b) 203.9 A; (c) 84.6 kVA.]

14 A coil of resistance 1.5 kΩ and 0.25 H inductance is connected in parallel with a variable capacitance across a 10 V, 8 kHz supply. Calculate (a) the capacitance of the capacitor when the supply current is a minimum; (b) the dynamic resistance, and (c) the supply current. [(a) 1561 pF; (b) 106.8 kΩ; (c) 93.66 μA.]

15 A single phase motor takes 30 A at a power factor of 0.65 lagging from a 240 V, 50 Hz supply. Determine (a) the current taken by the capacitor connected in parallel to correct the power factor to unity; and (b) the value of the supply current after power factor correction. [(a) 22.80 A; (b) 19.5 A.]

16 A motor has an output of 6 kW, an efficiency of 75% and a power factor of

0.64 lagging when operated from a 250 V, 60 Hz supply. It is required to raise the power factor to 0.925 lagging by connecting a capacitor in parallel with the motor. Determine (a) the current taken by the motor; (b) the supply current after power factor correction; (c) the current taken by the capacitor; (d) the capacitance of the capacitor and (e) the kvar rating of the capacitor.

[(a) 50 A; (b) 34.59 A; (c) 25.28 A; (d) 268.2 μF; (e) 6.32 kvar.]

17 A supply of 250 V, 80 Hz is connected across an inductive load and the power consumed is 2 kW, when the supply current is 10 A. Determine the resistance and inductance of the circuit. What value of capacitance connected in parallel with the load is needed to improve the overall power factor to unity?

[$R = 20\ \Omega$, $L = 29.84$ mH; $C = 47.75\ \mu$F]

18 A 200 V, 50 Hz single-phase supply feeds the following loads: (i) fluorescent lamps taking a current of 8 A at a power factor of 0.9 leading; (ii) incandescent lamps taking a current of 6 A at unity power factor; (iii) a motor taking a current of 12 A at a power factor of 0.65 lagging. Determine the total current taken from the supply and the overall power factor. Find also the value of a static capacitor connected in parallel with the loads to improve the overall power factor to 0.98 lagging. [21.74 A; 0.966 lagging; 21.68 μF]

4 Three-phase systems

A. MAIN POINTS CONCERNED WITH THREE-PHASE SYSTEMS

1 Generation, transmission and distribution of electricity via the National Grid system is accomplished by three-phase alternating currents.

2 The voltage induced by a single coil when rotated in a uniform magnetic field is shown in *Fig 1* and is known as a **single-phase voltage.** Most consumers are

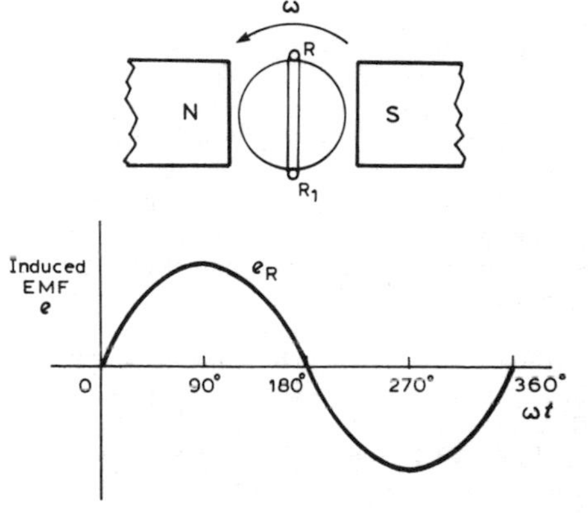

Fig 1

fed by means of a single-phase a.c. supply. Two wires are used, one called the live conductor (usually coloured red) and the other is called the neutral conductor (usually coloured black). The neutral is usually connected via protective gear to earth, the earth wire being coloured green. The standard voltage for a single-phase a.c. supply is 240 V. The majority of single-phase supplies are obtained by connection to a three-phase supply (see *Fig 5*).

3 A **three-phase supply** is generated when three coils are placed 120° apart and the whole rotated in a uniform magnetic field as shown in *Fig 2(a).* The result is three independent supplies of equal voltages which are each displaced by 120° from each other as shown in *Fig 2(b).*

4 (i) The convention adopted to identify each of the phase voltages is: R-red, Y-yellow, and B-blue, as shown in *Fig 2.*

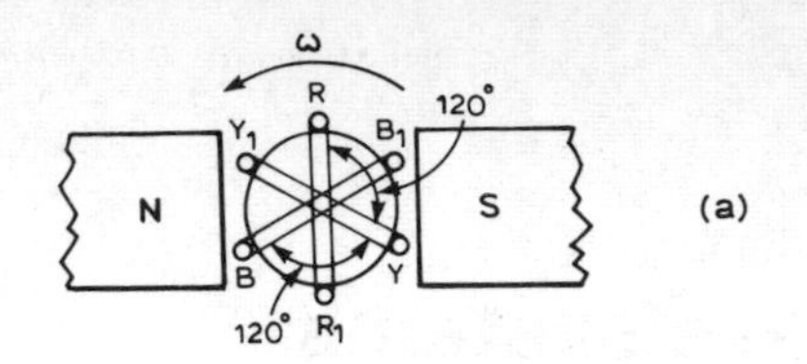

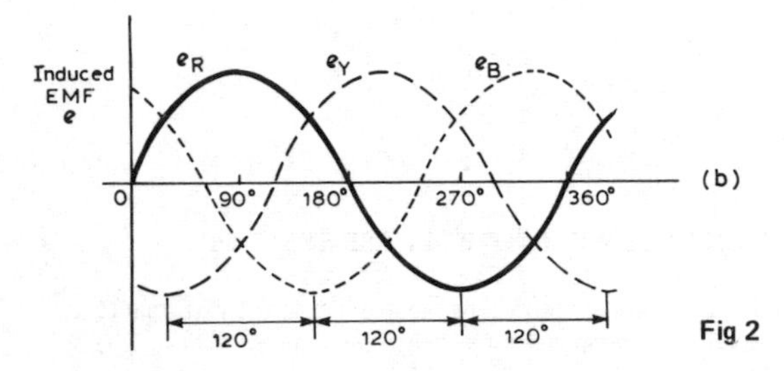

Fig 2

(ii) The **phase-sequence** is given by the sequence in which the conductors pass the point initially taken by the red conductor. The national standard phase sequence is R, Y, B.

5 A three-phase a.c. supply is carried by three conductors, called **'lines'** which are coloured red, yellow and blue. The currents in these conductors are known as line currents (I_L) and the p.d.'s between them are known as line voltages (V_L). A fourth conductor, called the **neutral** (coloured black, and connected through protective devices to earth) is often used with a three-phase supply.

6 If the three-phase windings shown in *Fig 2* are kept independent then six wires are needed to connect a supply source (such as a generator) to a load (such as motor). To reduce the number of wires it is usual to interconnect the three phases. There are two ways in which this can be done, these being: (a) a star connection, and (b) a delta, or mesh, connection. Sources of three-phase supplies, i.e. alternators, are usually connected in star, whereas three-phase transformer windings, motors and other loads may be connected either in star or delta.

7 (i) A **star-connected load** is shown in *Fig 3* where the three line conductors are each connected to a load and the outlets from the loads are joined together at N to form what is termed the **neutral point** or the **star point**.

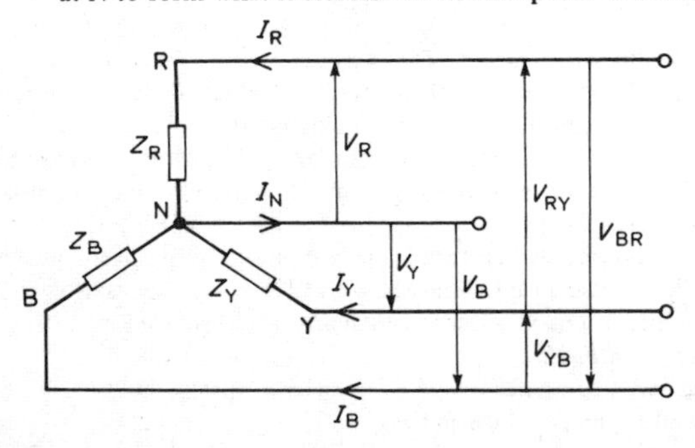

Fig 3

(ii) The voltages, V_R, V_Y and V_B are called **phase voltages** or line to neutral voltages. Phase voltages are generally denoted by V_p.

(iii) The voltages, V_{RY}, V_{YB} and V_{BR} are called **line voltages.**

(iv) From *Fig 3* it can be seen that the phase currents (generally denoted by I_p) are equal to their respective line currents I_R, I_Y and I_B, i.e. for a star connection:

$$I_L = I_p$$

(v) For a balanced system: $I_R = I_Y = I_B$, $V_R = V_Y = V_B$
$V_{RY} = V_{YB} = V_{BR}$, $Z_R = Z_Y = Z_B$
and the current in the neutral conductor, $I_N = 0$.
When a star connected system is balanced, then the neutral conductor is unnecessary and is often omitted.

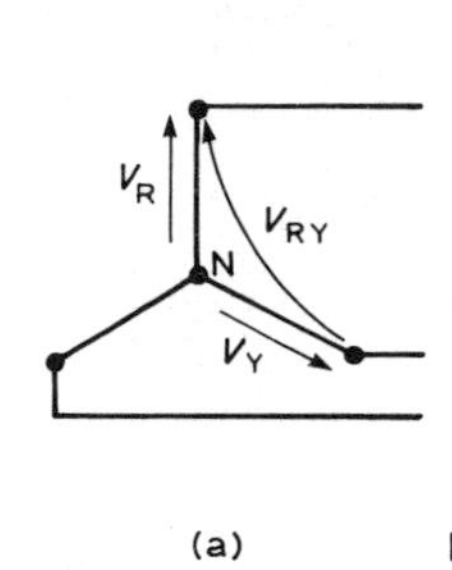

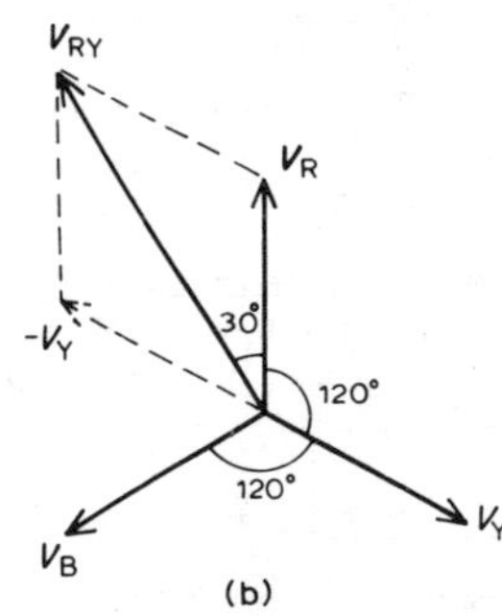

Fig 4

(vi) The line voltage, V_{RY}, shown in *Fig 4(a)* is given by $V_{RY} = V_R - V_Y$. (V_Y is negative since it is in the opposite direction to V_{RY}.) In the phasor diagram of *Fig 4(b)*, phasor V_Y is reversed (shown by the broken line) and then added phasorially to V_R (i.e. $V_{RY} = V_R + (-V_Y)$). By trigonometry, or by measurement, $V_{RY} = \sqrt{3} V_R$, i.e. for a balanced star connection:

$$V_L = \sqrt{3}\ V_p$$

(See *Problem 6* for a complete phasor diagram of a star-connected system.)

(vii) The star connection of the three phases of a supply, together with a neutral conductor, allows the use of two voltages–the phase voltage and the line voltage. A 4-wire system is also used when the load is not balanced. The standard electricity supply to consumers in Great Britain is 415/240 V, 50 Hz, 3-phase, 4-wire alternating current, and a diagram of connections is shown in *Fig 5*.

8 (i) **A delta (or mesh) connected load** is shown in *Fig 6* where the end of one load is connected to the start of the next load.

(ii) From *Fig 6*, it can be seen that the line voltages V_{RY}, V_{YB} and V_{BR} are the respective phase voltages, i.e. for a delta connection:

$$V_L = V_p$$

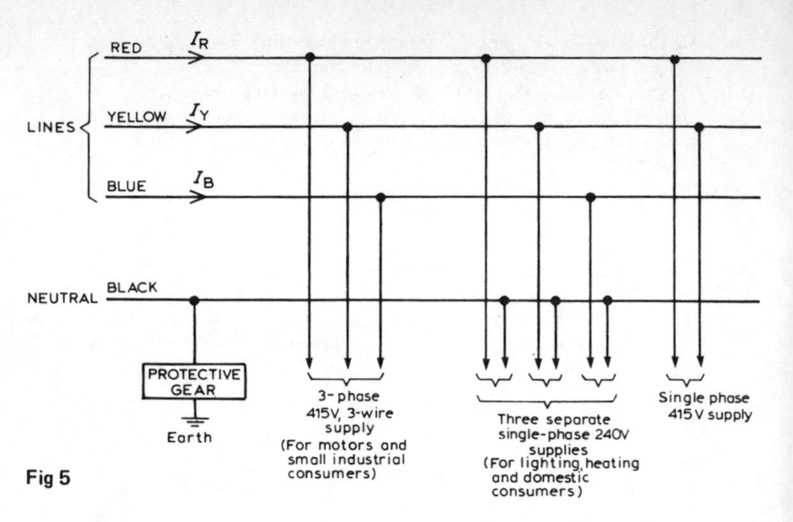

Fig 5

(iii) Using Kirchhoff's current law in *Fig 6*, $I_R = I_{RY} - I_{BR} = I_{RY} + (-I_{BR})$. From the phasor diagram shown in *Fig 7*, by trigonometry or by measurement, $I_R = \sqrt{3}\, I_{RY}$, i.e. for a delta connection:

$$I_L = \sqrt{3}\, I_p$$

9 The power dissipated in a three-phase load is given by the sum of the power dissipated in each phase. If a load is balanced then the total power P is given by:

$P = 3 \times$ power consumed by one phase.

The power consumed in one phase $= I_p^2 R_p$ or $V_p I_p \cos\phi$ (where ϕ is the phase angle between V_p and I_p).

For a star connection $V_p = \dfrac{V_L}{\sqrt{3}}$ and $I_p = I_L$ hence $P = 3\left(\dfrac{V_L}{\sqrt{3}}\right) I_L \cos\phi$

$$= \sqrt{3}\, V_L I_L \cos\phi.$$

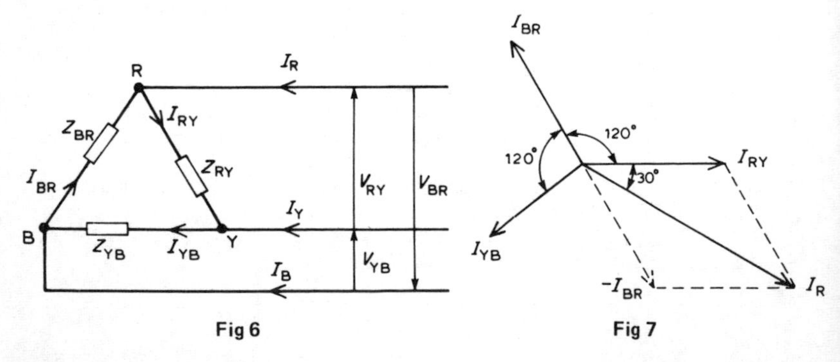

Fig 6

Fig 7

For a delta connection, $V_p = V_L$ and $I_p = \dfrac{I_L}{\sqrt{3}}$ hence $P = 3V_L\left(\dfrac{I_L}{\sqrt{3}}\right)\cos\phi$

$$= \sqrt{3}\, V_L I_L \cos\phi.$$

Hence for either a star or a delta balanced connection the total power P is given by:

$P = \sqrt{3}\, V_L I_L \cos\phi$ **watts** or $P = 3I_p{}^2 R_p$ **watts.**

Total volt-amperes, $S = \sqrt{3}\, V_L I_L$ **volt-amperes.**

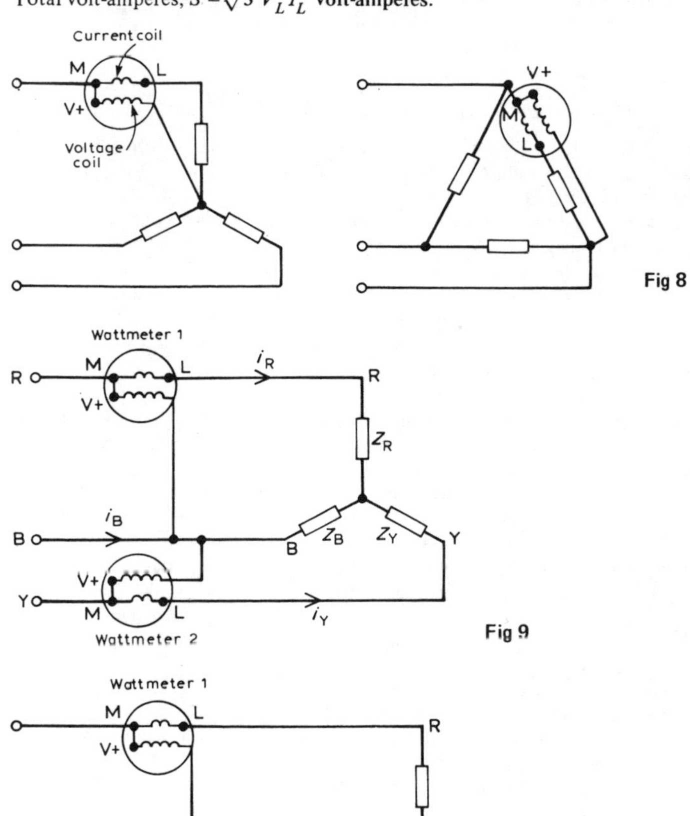

Fig 8

Fig 9

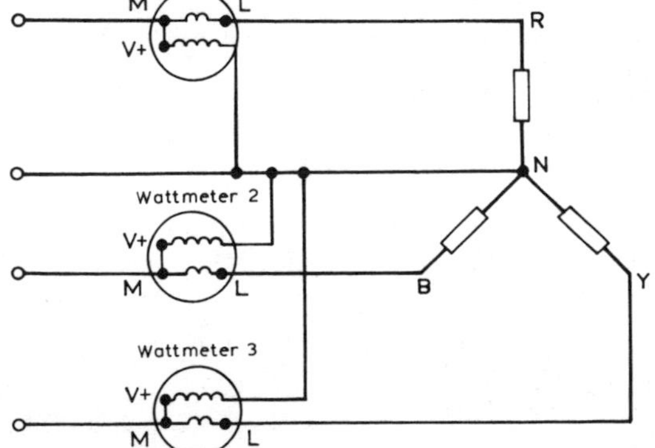

Fig 10

10 Power in three-phase loads may be measured by the following methods:

(i) **One-wattmeter method for a balanced load.**
Wattmeter connections for both star and delta are shown in *Fig 8*.
Total power = 3 × wattmeter reading

(ii) **Two-wattmeter method for balanced or unbalanced loads.**
A connection diagram for this method is shown in *Fig 9* for a star-connected load. Similar connections are made for a delta-connected load.
Total power = sum of wattmeter readings = $P_1 + P_2$.
The power factor may be determined from:

$$\tan \phi = \sqrt{3}\,\frac{(P_1 + P_2)}{(P_1 + P_2)}$$ (see *Problems 12 and 15 to 18*).

It is possible, depending on the load power factor, for one wattmeter to have to be 'reversed' to obtain a reading. In this case it is taken as a negative reading (see *Problem 17*).

(iii) **Three-wattmeter method for a three-phase, 4-wire system for balanced and unbalanced loads.** (see *Fig 10*)
Total power = $P_1 + P_2 + P_3$

11 (i) Loads connected in delta dissipate three times more power than when connected in star to the same supply.

(ii) For the same power, the phase currents must be the same for both delta and star connections (since power = $3I_p^2 R_p$), hence the line current in the delta-connected system is greater than the line current in the corresponding star-connected system. To achieve the same phase current in a star-connected system as in a delta-connected system, the line voltage in the star system is $\sqrt{3}$ times the line voltage in the delta system.

Thus for a given power transfer, a delta system is associated with larger line currents (and thus larger conductor cross-sectional area) and a star system is associated with a larger line voltage (and thus greater insulation).

12 **Advantages of three-phase systems** over single-phase supplies include:

(i) For a given amount of power transmitted through a system, the three-phase system requires conductors with a smaller cross-sectional area. This means a saving of copper (or aluminium) and thus the original installation costs are less.

(ii) Two voltages are available (see para. 7).

(iii) Three-phase motors are very robust, relatively cheap, generally smaller, have self-starting properties, provide a steadier output and require little maintenance compared with single-phase motors.

B. WORKED PROBLEMS ON THREE-PHASE SYSTEMS

Problem 1 Three loads, each of resistance 30 Ω, are connected in star to a 415 V, 3-phase supply. Determine (a) the system phase voltage; (b) the phase current and (c) the line current.

A '415 V, 3-phase supply' means that 415 V is the line voltage, V_L.

(a) For a star connection, $V_L = \sqrt{3}\, V_p$

Hence phase voltage, $V_p = \frac{V_L}{\sqrt{3}} = \frac{415}{\sqrt{3}} =$ **239.6 V or 240 V** correct to 3 significant figures.

(b) Phase current, $I_p = \frac{V_p}{R_p} = \frac{240}{30} = \mathbf{8\ A}$

(c) For a star connection, $I_p = I_L$
Hence the line current, $I_L = \mathbf{8\ A}$

Problem 2 A star-connected load consists of three identical coils each of resistance 30 Ω and inductance 127.3 mH. If the line current is 5.08 A, calculate the line voltage if the supply frequency is 50 Hz.

Inductive reactance $X_L = 2\pi fL = 2\pi(50)(127.3 \times 10^{-3}) = 40\ \Omega$
Impedance of each phase $Z_p = \sqrt{(R^2 + X_L^2)} = \sqrt{(30^2 + 40^2)} = 50\ \Omega$

For a star connection $I_L = I_p = \frac{V_p}{Z_p}$

Hence phase voltage $V_p = I_p Z_p = (5.08)(50) = 254$ V
Line voltage $V_L = \sqrt{3} V_p = \sqrt{3}\,(254) = \mathbf{440\ V}$

Problem 3 The three coils in *Problem 2* are now connected in delta to the 440 V, 50 Hz, 3-phase supply. Determine (a) the phase current and (b) the line current.

Phase impedance, $Z_p = 50\ \Omega$ (as above) and for a delta connection $V_p = V_L$

(a) Phase current, $I_p = \frac{V_p}{Z_p} = \frac{V_L}{Z_p} = \frac{440}{50} = \mathbf{8.8\ A}$

(b) For a delta connection, $I_L = \sqrt{3}\, I_p = \sqrt{3}\,(8.8) = \mathbf{15.24\ A}$

Thus when the load is connected in delta, three times the line current is taken from the supply than is taken if connected in star.

Problem 4 Three identical capacitors are connected in delta to a 415 V, 50 Hz, 3-phase supply. If the line current is 15 A, determine the capacitance of each of the capacitors.

For a delta connection $I_L = \sqrt{3}\, I_p$

Hence phase current $I_p = \frac{I_L}{\sqrt{3}} = \frac{15}{\sqrt{3}} = 8.66$ A

Capacitive reactance per phase, $X_C = \frac{V_p}{I_p} = \frac{V_L}{I_p}$ (since for a delta connection $V_L = V_p$).

Hence $X_C = \frac{415}{8.66} = 47.92\ \Omega$

$X_C = \frac{1}{2\pi fC}$, from which capacitance, $C = \frac{1}{2\pi f X_C} = \frac{1}{2\pi(50)(47.92)}$ F
$= \mathbf{66.43\ \mu F}$

Problem 5 Three coils each having resistance 3 Ω and inductive reactance 4 Ω are connected (i) in star and (ii) in delta to a 415 V, 3-phase supply. Calculate for each connection (a) the line and phase voltages and (b) the phase and line currents.

(i) **For a star connection:** $I_L = I_p$ and $V_L = \sqrt{3}\, V_p$

(a) A 415 V, 3-phase supply means that the line voltage, V_L = **415 V**

Phase voltage, $V_p = \dfrac{V_L}{\sqrt{3}} = \dfrac{415}{\sqrt{3}}$ = **240 V**

(b) Impedance per phase, $Z_p = \sqrt{(R^2 + X_L{}^2)} = \sqrt{(3^2 + 4^2)} = 5\ \Omega$

Phase current, $I_p = \dfrac{V_p}{Z_p} = \dfrac{240}{5}$ = **48 A**

Line current, $I_L = I_p$ = **48 A**

(ii) **For a delta connection:** $V_L = V_p$ and $I_L = \sqrt{3}\, I_p$

(a) Line voltage, V_L = **415 V**

Phase voltage, $V_p = V_L$ = **415 V**

(b) Phase current, $I_p = \dfrac{V_p}{Z_p} = \dfrac{415}{5}$ = **83 A**

Line current, $I_L = \sqrt{3}\, I_p = \sqrt{3}(83)$ = **144 A**

Problem 6 A balanced, three-wire, star-connected, 3-phase load has a phase voltage of 240 V, a line current of 5 A and a lagging power factor of 0.966. Draw the complete phasor diagram.

The phasor diagram is shown in *Fig 11*.

Procedure to construct the phasor diagram:

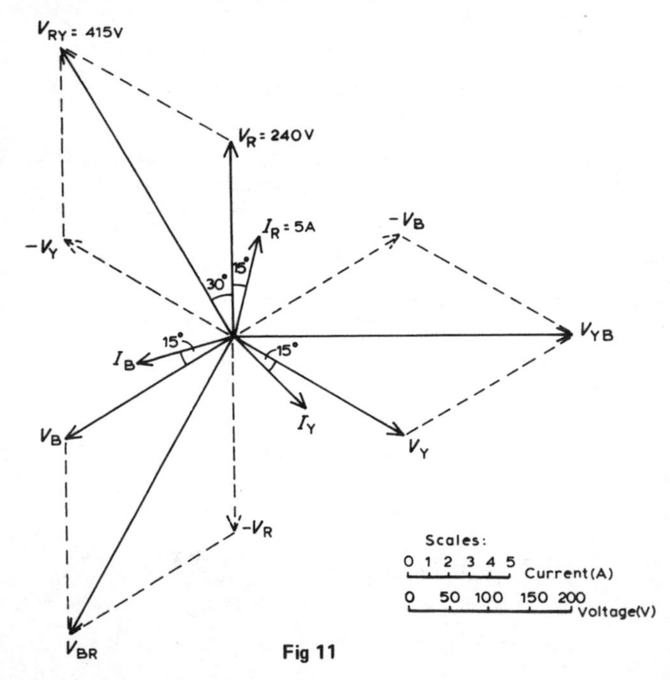

Fig 11

(i) Draw $V_R = V_Y = V_B = 240$ V and spaced 120° apart. (Note that V_R is shown vertically upwards–this however is immaterial for it may be drawn in any direction.)

(ii) Power factor $= \cos\phi = 0.966$ lagging. Hence the load phase angle is given by arccos 0.966, i.e. 15° lagging. Hence $I_R = I_Y = I_B = 5$ A, lagging V_R, V_Y and V_B respectively by 15°.

(iii) $V_{RY} = V_R - V_Y$ (phasorially). Hence V_Y is reversed and added phasorially to V_R. By measurement, $V_{RY} = 415$ V (i.e. $\sqrt{3}(240)$) and leads V_R by 30°. Similarly, $V_{YB} = V_Y - V_B$ and $V_{BR} = V_B - V_R$.

Problem 7 A 415 V, 3 phase, 4 wire, star-connected system supplies three resistive loads as shown in *Fig 12*. Determine (a) the current in each line and (b) the current in the neutral conductor.

(a) For a star-connected system $V_L = \sqrt{3} V_P$

Hence $\quad V_P = \dfrac{V_L}{\sqrt{3}} = \dfrac{415}{\sqrt{3}} = 240\ V$

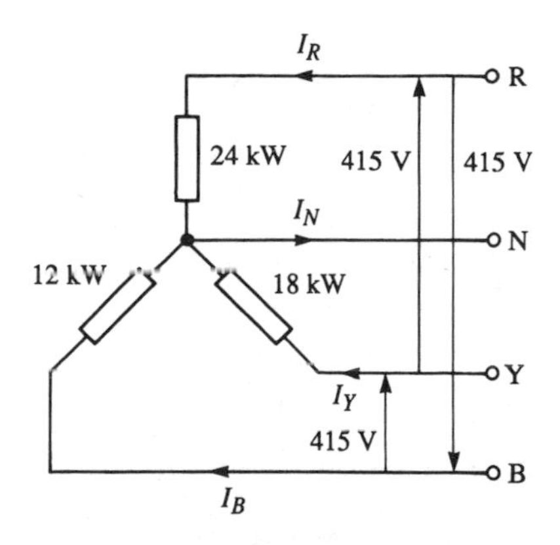

Fig 12

Since current $I = \dfrac{\text{Power } P}{\text{Voltage } V}$ for a resistive load

then $\quad I_R = \dfrac{P_R}{V_R} = \dfrac{24000}{240} = \mathbf{100\ A}$

$I_Y = \dfrac{P_Y}{V_Y} = \dfrac{18000}{240} = \mathbf{75\ A}$

and $\quad I_B = \dfrac{P_B}{V_B} = \dfrac{12000}{240} = \mathbf{50\ A}$

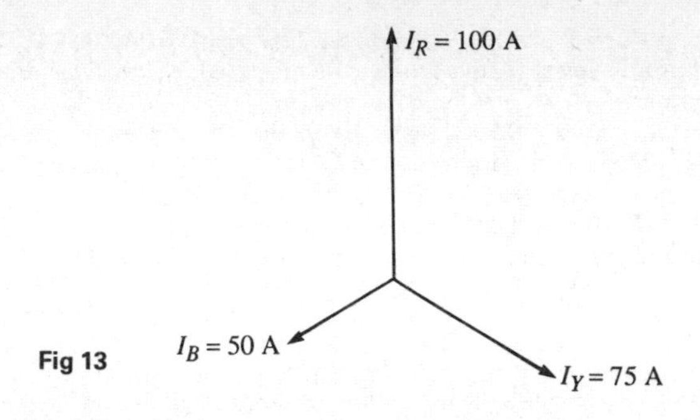

Fig 13

(b) The three line currents are shown in the phasor diagram of *Fig 13*. Since each load is resistive the currents are in phase with the phase voltages and are hence mutually displaced by 120°. The current in the neutral conductor is given by:

$I_N = I_R + I_Y + I_B$ phasorially.

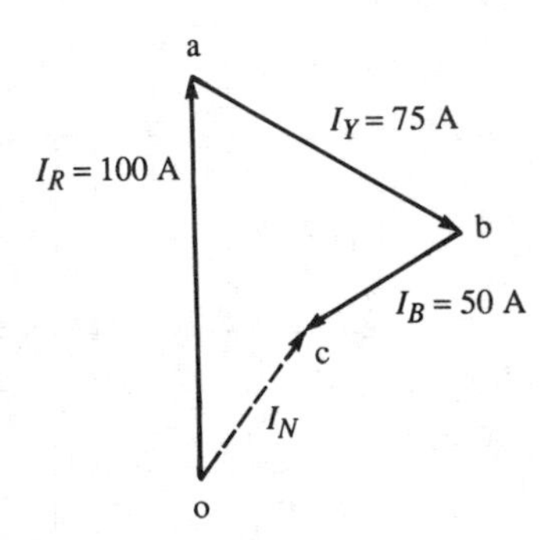

Fig 14

Fig 14 shows the three line currents added phasorially.
oa represents I_R in magnitude and direction.
From the nose of oa, ab is drawn representing I_Y in magnitude and direction.
From the nose of ab, bc is drawn representing I_B in magnitude and direction.
oc represents the resultant, I_N. By measurement, $I_N =$ **43 A**
(Alternatively, by calculation, considering I_R at 90°, I_B at 210° and I_Y at 330°:
Total horizontal component $= 100\cos 90° + 75\cos 330° + 50\cos 210° = 21.65$
Total vertical component $= 100\sin 90° + 75\sin 330° + 50\sin 210° = 37.50$
Hence magnitude of $I_N = \sqrt{(21.65^2 + 37.50^2)} =$ **43.3 A**)

Problem 8 Three 12 Ω resistors are connected in star to a 415 V, 3-phase supply. Determine the total power dissipated by the resistors.

Power dissipated, $P = \sqrt{3}\, V_L I_L \cos\phi$ or $P = 3 I_p^2 R_p$.

Line voltage, $V_L = 415$ V and phase voltage $V_p = \dfrac{415}{\sqrt{3}} = 240$ V

(since the resistors are star-connected)

Phase current, $I_p = \dfrac{V_p}{Z_p} = \dfrac{V_p}{R_p} = \dfrac{240}{12} = 20$ A

For a star connection, $I_L = I_p = 20$ A

For a purely resistive load, the power factor = $\cos\phi = 1$

Hence power $P = \sqrt{3}\, V_L I_L \cos\phi = \sqrt{3}\,(415)(20)(1) =$ **14.4 kW**

or power $P = 3 I_p^2 R_p = 3(20)^2(12) =$ **14.4 kW**

Problem 9 The input power to a 3-phase a.c. motor is measured as 5 kW. If the voltage and current to the motor are 400 V and 8.6 A respectively, determine the power factor of the system.

Power, $P = 5000$ W; Line voltage $V_L = 400$ V; Line current, $I_L = 8.6$ A

Power, $P = \sqrt{3}\, V_L I_L \cos\phi$

Hence power factor = $\cos\phi = \dfrac{P}{\sqrt{3}\, V_L I_L} = \dfrac{5000}{\sqrt{3}(400)(8.6)} =$ **0.839**

Problem 10 Three identical coils, each of resistance 10 Ω and inductance 42 mH are connected (a) in star and (b) in delta to a 415 V, 50 Hz, 3-phase supply. Determine the total power dissipated in each case.

(a) **Star-connection**

Inductive reactance $X_L = 2\pi f L = 2\pi(50)(42 \times 10^{-3}) = 13.19$ Ω

Phase impedance $Z_p = \sqrt{(R^2 + X_L^2)} = \sqrt{(10^2 + 13.19^2)} = 16.55$ Ω

Line voltage $V_L = 415$ V and phase voltage, $V_p = \dfrac{V_L}{\sqrt{3}} = \dfrac{415}{\sqrt{3}} = 240$ V

Phase current $I_p = \dfrac{V_p}{Z_p} = \dfrac{240}{16.55} = 14.50$ A

Line current $I_L = I_p = 14.50$ A

Power factor = $\cos\phi = \dfrac{R_p}{Z_p} = \dfrac{10}{16.55} = 0.6042$ lagging

Power dissipated $P = \sqrt{3}\, V_L I_L \cos\phi = \sqrt{3}(415)(14.50)(0.6042) =$ **6.3 kW**

(Alternatively $P = 3 I_p^2 R_p = 3(14.50)^2(10) =$ **6.3 kW**)

(b) **Delta-connection**

$V_L = V_p = 415$ V; $Z_p = 16.55\ \Omega$; $\cos\phi = 0.6042$ lagging (from above).

Phase current $I_p = \dfrac{V_p}{Z_p} = \dfrac{415}{16.55} = 25.08$ A

Line current $I_L = \sqrt{3} I_p = \sqrt{3}(25.08) = 43.44$ A
Power dissipated $P = \sqrt{3}\, V_L I_L \cos\phi = \sqrt{3}(415)(43.44)(0.6042) = \mathbf{18.87\ kW}$
(Alternatively $P = 3 I_p^2 R_p = 3(25.08)^2(10) = \mathbf{18.87\ kW}$)

Hence loads connected in delta dissipate three times the power than when connected in star and also take a line current three times greater.

Problem 11 A 415 V, 3-phase a.c. motor has a power output of 12.75 kW and operates at a power factor of 0.77 lagging and with an efficiency of 85%. If the motor is delta-connected, determine (a) the power input; (b) the line current and (c) the phase current.

(a) Efficiency $= \dfrac{\text{power output}}{\text{power input}}$ Hence $\dfrac{85}{100} = \dfrac{12\,750}{\text{power input}}$

from which, power input $= \dfrac{12\,750 \times 100}{85} = 15\,000$ W or **15 kW**

(b) Power, $P = \sqrt{3}\, V_L I_L \cos\phi$, hence line current, $I_L = \dfrac{P}{\sqrt{3}\, V_L \cos\phi}$

$= \dfrac{15\,000}{\sqrt{3}(415)(0.77)} = \mathbf{27.10\ A}$

(c) For a delta connection, $I_L = \sqrt{3} I_p$. Hence phase current, $I_p = \dfrac{I_L}{\sqrt{3}} = \dfrac{27.10}{\sqrt{3}}$
$= \mathbf{15.65\ A}$

Problem 12 (a) Show that the total power in a 3-phase, 3-wire system using the two-wattmeter method of measurement is given by the sum of the wattmeter readings. Draw a connection diagram.
(b) Draw a phasor diagram for the two-wattmeter method for a balanced load.
(c) Use the phasor diagram of part (b) to derive a formula from which the power factor of a 3-phase system may be determined using only the wattmeter readings.

(a) A connection diagram for the two-wattmeter method of power measurement is shown in *Fig 15* for a star-connected load.
Total instantaneous power, $p = e_R i_R + e_Y i_Y + e_B i_B$ and in any 3 phase system $i_R + i_Y + i_B = 0$. Hence $i_B = -i_R - i_Y$.
Thus, $p = e_R i_R + e_Y i_Y + e_B(-i_R - i_Y)$
$= (e_R - e_B) i_R + (e_Y - e_B) i_Y$
However, $(e_R - e_B)$ is the p.d. across wattmeter 1 in *Fig 15* and $(e_Y - e_B)$ is the p.d. across wattmeter 2.
Hence total instantaneous power
$p =$ (wattmeter 1 reading) + (wattmeter 2 reading) $= p_1 + p_2$.

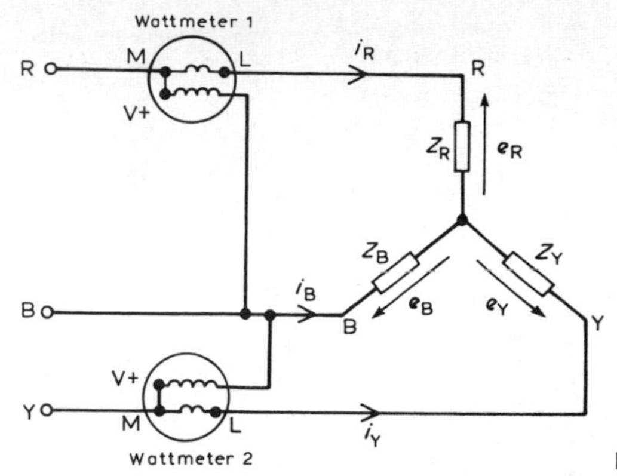

Fig 15

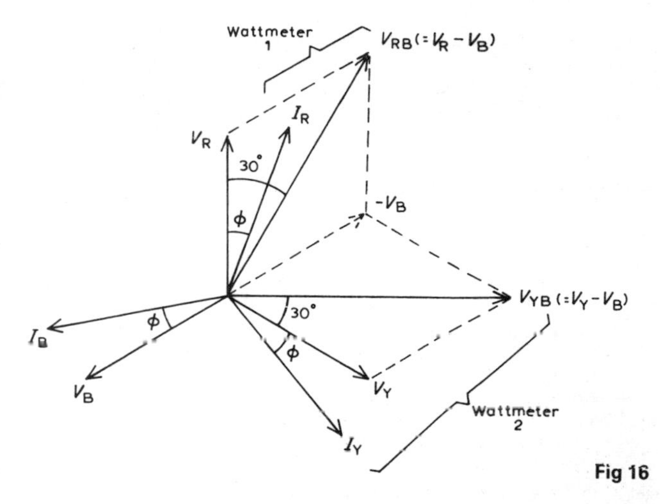

Fig 16

The moving systems of the wattmeters are unable to follow the variations which take place at normal frequencies and they indicate the mean power taken over a cycle. Hence the total power, $P = P_1 + P_2$ for balanced or unbalanced loads.

(b) The phasor diagram for the two-wattmeter method for a balanced load having a lagging current is shown in *Fig 16*, where $V_{RB} = V_R - V_B$ and $V_{YB} = V_Y - V_B$ (phasorially).

(c) Wattmeter 1 reads $V_{RB}I_R \cos(30° - \phi) = P_1$
Wattmeter 2 reads $V_{YB}I_Y \cos(30° + \phi) = P_2$

$$\frac{P_1}{P_2} = \frac{V_{RB}I_R \cos(30° - \phi)}{V_{YB}I_Y \cos(30° + \phi)} = \frac{\cos(30° - \phi)}{\cos(30° + \phi)},$$

since $I_R = I_Y$ and $V_{RB} = V_{YB}$ for a balanced load.

Hence $\dfrac{P_1}{P_2} = \dfrac{\cos 30° \cos \phi + \sin 30° \sin \phi}{\cos 30° \cos \phi - \sin 30° \sin \rho}$ (from compound angle formulae)

Dividing throughout by cos 30° cos ϕ gives:

$$\frac{P_1}{P_2} = \frac{1+\tan 30° \tan \phi}{1-\tan 30° \tan \phi} = \frac{1+\frac{1}{\sqrt{3}} \tan \phi}{1-\frac{1}{\sqrt{3}} \tan \phi}, \text{ (since } \frac{\sin \phi}{\cos \phi} = \tan \phi)$$

Cross-multiplying gives: $P_1 - \dfrac{P_1}{\sqrt{3}} \tan \phi = P_2 + \dfrac{P_2}{\sqrt{3}} \tan \phi.$

$$\text{Hence } P_1 - P_2 = (P_1 + P_2) \frac{\tan \phi}{\sqrt{3}}$$

from which $\mathbf{\tan \phi = \sqrt{3} \left(\dfrac{P_1 - P_2}{P_1 + P_2}\right)}$

ϕ, cos ϕ and thus power factor can be determined from this formula.

Problem 13 A 400 V, 3-phase star connected alternator supplies a delta-connected load, each phase of which has a resistance of 30 Ω and inductive reactance 40 Ω. Calculate (a) the current supplied by the alternator and (b) the output power and the kVA of the alternator, neglecting losses in the line between the alternator and load.

A circuit diagram of the alternator and load is shown in *Fig 17.*

(a) Considering the load: Phase current, $I_p = \dfrac{V_p}{Z_p}$

$V_p = V_L$ for a delta connection. Hence $V_p = 400$ V
Phase impedance, $Z_p = \sqrt{(R_p{}^2 + X_L{}^2)} = \sqrt{(30^2 + 40^2)} = 50\ \Omega$

Hence $I_p = \dfrac{V_p}{Z_p} = \dfrac{400}{50} = 8$ A

For a delta-connection, line current, $I_L = \sqrt{3} I_p = \sqrt{3}(8) = 13.86$ A
Hence 13.86 A is the current supplied by the alternator.

(b) Alternator output power is equal to the power dissipated by the load,

i.e. $P = \sqrt{3}\, V_L I_L \cos \phi$, where $\cos \phi = \dfrac{R_p}{Z_p} = \dfrac{30}{50} = 0.6$

Hence $P = \sqrt{3}(400)(13.86)(0.6)$
$= \mathbf{5.76\ kW}$

Alternator output kVA, $S = \sqrt{3}\, V_L I_L$
$= \sqrt{3}(400)(13.86)$
$= \mathbf{9.60\ kVA}$

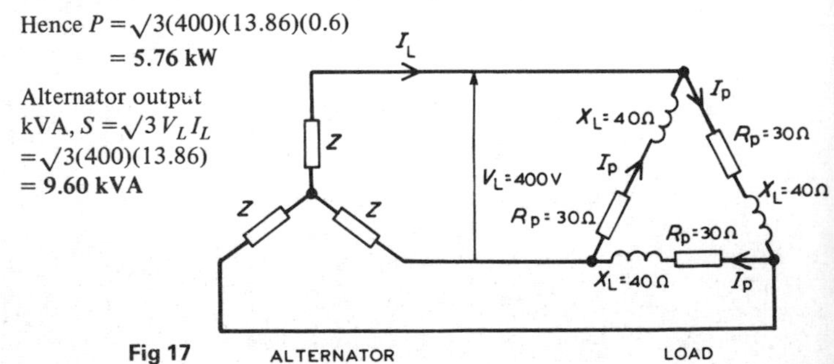

Fig 17

Problem 14 Each phase of a delta-connected load comprises a resistance of 30 Ω and an 80 μF capacitor in series. The load is connected to a 400 V, 50 Hz, 3-phase supply. Calculate (a) the phase current; (b) the line current; (c) the total power dissipated and (d) the kVA rating of the load. Draw the complete phasor diagram for the load.

(a) Capacitive reactance, $X_C = \dfrac{1}{2\pi fC} = \dfrac{1}{2\pi(50)(80 \times 10^{-6})} = 39.79\ \Omega$

Phase impedance, $Z_p = \sqrt{(R_p{}^2 + X_C{}^2)} = \sqrt{(30^2 + 39.79^2)} = 49.83\ \Omega$

Power factor $= \cos\phi = \dfrac{R_p}{Z_p} = \dfrac{30}{49.83} = 0.602$. Hence $\phi = \arccos 0.602$
$= 52°\ 59'$ leading.

Phase current, $I_p = \dfrac{V_p}{Z_p}$ and $V_p = V_L$ for a delta connection.

Hence $I_p = \dfrac{400}{49.83} = \mathbf{8.027\ A}$

(b) Line current $I_L = \sqrt{3}\, I_p$ for a delta-connection.
Hence $I_L = \sqrt{3}\,(8.027) = \mathbf{13.90\ A}$

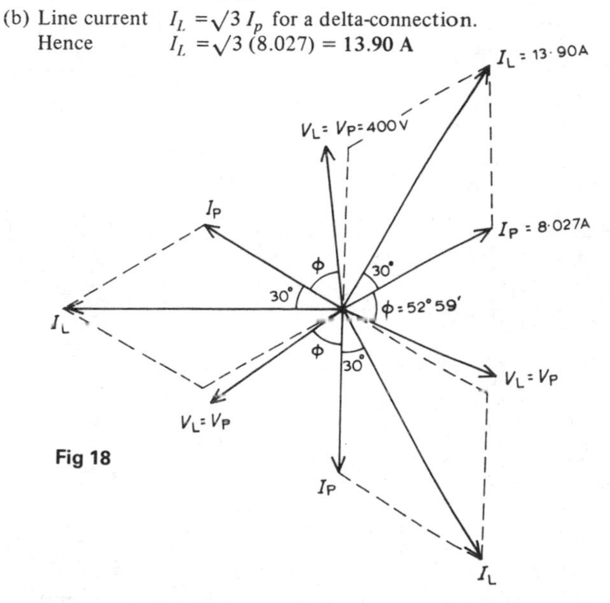

Fig 18

(c) Total power dissipated, $P = \sqrt{3}\, V_L I_L \cos\phi = \sqrt{3}(400)(13.90)(0.602)$
$= \mathbf{5.797\ kW}$

(d) Total kVA, $S = \sqrt{3}\, V_L I_L = \sqrt{3}(400)(13.90) = \mathbf{9.630\ kVA}$
The phasor diagram for the load is shown in *Fig 18*.

Problem 15 Two wattmeters are connected to measure the input power to a balanced 3-phase load by the two-wattmeter method. If the instrument readings are 8 kW and 4 kW, determine (a) the total power input and (b) the load power factor.

With reference to para. 10(ii):

(a) Total input power $P = P_1 + P_2 = 8+4 =$ **12 kW**

(b) $\tan\phi = \sqrt{3}\left(\frac{P_1 - P_2}{P_1 + P_2}\right) = \sqrt{3}\left(\frac{8-4}{8+4}\right) = \sqrt{3}\left(\frac{4}{12}\right) = \sqrt{3}\left(\frac{1}{3}\right) = \frac{1}{\sqrt{3}}$

Hence $\phi = \arctan\frac{1}{\sqrt{3}} = 30°$

Power factor $= \cos\phi = \cos 30° =$ **0.866**

Problem 16 Two wattmeters connected to a 3-phase motor indicate the total power input to be 12 kW. The power factor is 0.6. Determine the readings of each wattmeter.

If the two wattmeters indicate P_1 and P_2 respectively then $P_1 + P_2 = 12$ kW (1)

$\tan\phi = \sqrt{3}\left(\frac{P_1 - P_2}{P_1 + P_2}\right)$ and power factor $= 0.6 = \cos\phi$

Angle $\phi = \arccos 0.6 = 53°\ 8'$ and $\tan 53°\ 8' = 1.3333$

Hence $1.3333 = \frac{\sqrt{3}(P_1 - P_2)}{12}$ from which $P_1 - P_2 = \frac{12(1.3333)}{\sqrt{3}} = 9.237$ kW (2)

Adding equations (1) and (2) gives: $2P_1 = 21.237$, i.e. $P_1 = \frac{21.237}{2} = 10.62$ kW

Hence **wattmeter 1 reads 10.62 kW**

From equation (1), **wattmeter 2 reads (12–10.62) = 1.38 kW**

Problem 17 Two wattmeters indicate 10 kW and 3 kW respectively when connected to measure the input power to a 3-phase balanced load, the reverse switch being operated on the meter indicating the 3 kW reading. Determine (a) the input power and (b) the load power factor.

Since the reversing switch on the wattmeter had to be operated the 3 kW reading is taken as –3 kW.

(a) Total input power, $P = P_1 + P_2 = 10+(-3) =$ **7 kW**

(b) $\tan\phi = \sqrt{3}\left(\frac{P_1 - P_2}{P_1 + P_2}\right) = \sqrt{3}\left(\frac{10-(-3)}{10+(-3)}\right) = \sqrt{3}\left(\frac{13}{7}\right) = 3.2167$

Angle $\phi = \arctan 3.2167 = 72°\ 44'$.

Power factor $= \cos\phi = \cos 72°\ 44' =$ **0.297**

Problem 18 Three similar coils, each having a resistance of 8 Ω and an inductive reactance of 8 Ω are connected (a) in star and (b) in delta, across a 415 V, 3-phase supply. Calculate for each connection the readings on each of two wattmeters connected to measure the power by the two-wattmeter method.

(a) **Star connection**: $V_L = \sqrt{3}V_p$ and $I_L = I_p$

Phase voltage, $V_p = \frac{V_L}{\sqrt{3}} = \frac{415}{\sqrt{3}}$ and phase impedance, $Z_p = \sqrt{(R_p{}^2 + X_L^2)}$

$= \sqrt{(8^2 + 8^2)}$

$= 11.31\ \Omega$

Hence phase current, $I_p = \frac{V_p}{Z_p} = \frac{415/\sqrt{3}}{11.31} = 21.18$ A

Total power $P = 3I_p^2R_p = 3(21.18)^2(8) = 10\,766$ W

If wattmeter readings are P_1 and P_2 then $P_1 + P_2 = 10\,766$ (1)

Since $R_p = 8\ \Omega$ and $X_L = 8\ \Omega$, then phase angle $\phi = 45°$ (from impedance triangle)

$$\tan\phi = \sqrt{3}\left(\frac{P_1 - P_2}{P_1 + P_2}\right), \text{ hence } \tan 45° = \frac{\sqrt{3}(P_1 - P_2)}{10\,766}$$

$$\text{from which } P_1 - P_2 = \frac{10\,766\,(1)}{\sqrt{3}} = 6216 \text{ W} \qquad (2)$$

Adding equations (1) and (2) gives: $2P_1 = 10\,766 + 6216 = 16\,982$ W

Hence $P_1 = 8491$ W

From equation (1), $P_2 = 10\,766 - 8491 = 2275$ W

When the coils are star-connected the wattmeter readings are thus 8.491 kW and 2.275 kW.

(b) **Delta connection:** $V_L = V_p$ and $I_L = \sqrt{3}I_p$

$$\text{Phase current, } I_p = \frac{V_p}{Z_p} = \frac{415}{11.31} = 36.69 \text{ A}$$

Total power $P = 3\,I_p^2R_p = 3(36.69)^2(8) = 32\,310$ W

Hence $P_1 + P_2 = 32\,310$ W (3)

$$\tan\phi = \sqrt{3}\left(\frac{P_1 - P_2}{P_1 + P_2}\right) \text{ thus } 1 = \frac{\sqrt{3}(P_1 - P_2)}{32\,310}$$

$$\text{from which } P_1 - P_2 = \frac{32\,310}{\sqrt{3}} = 18\,650 \text{ W} \qquad (4)$$

Adding equations (3) and (4) gives: $2P_1 = 50\,960$, from which $P_1 = 25\,480$ W

From equation (3), $P_2 = 32\,310 - 25\,480 = 6830$ W

When the coils are delta-connected the wattmeter readings are thus 25.48 kW and 6.83 kW.

C. FURTHER PROBLEMS ON THREE-PHASE SYSTEMS

SHORT ANSWER PROBLEMS

1 Explain briefly how a three-phase supply is generated.

2 State the national standard phase sequence for a three-phase supply.

3 State the two ways in which phases of a three-phase supply can be interconnected to reduce the number of conductors used compared with three, single-phase systems.

4 State the relationships between line and phase currents and line and phase voltages for a star-connected system.

5 When may the neutral conductor of a star-connected system be omitted?

6 State the relationships between line and phase currents and line and phase voltages for a delta-connected system.

7 What is the standard electricity supply to consumers in Great Britain?

8 State two formulae for determining the power dissipated in the load of a 3-phase balanced system.

9 By what methods may power be measured in a three-phase system?

10 State a formula from which power factor may be determined for a balanced system when using the two-wattmeter method of power measurement.

11 Loads connected in star dissipate .... the power dissipated when connected in delta and fed from the same supply.
12 Name three advantages of three-phase systems over single-phase systems.

MULTI-CHOICE PROBLEMS (answers on page 191)

Three loads, each of 10 Ω resistance, are connected in star to a 400 V, 3-phase supply. Determine the quantities stated in Problems 1 to 5, selecting the correct answers from the following list.

(a) $\frac{40}{\sqrt{3}}$ A; (b) $\sqrt{3}$ (16) kW; (c) $\frac{400}{\sqrt{3}}$ V; (d) $\sqrt{3}$ (40) A; (e) $\sqrt{3}$ (400) V; (f) 16 kW; (g) 400 V; (k) 48 kW; (i) 40 A.

1 Line voltage.
2 Phase voltage.
3 Phase current.
4 Line current.
5 Total power dissipated in the load.
6 Which of the following statements is false?
 (a) For the same power, loads connected in delta have a higher line voltage and a smaller line current than loads connected in star.
 (b) When using the two-wattmeter method of power measurement the power factor is unity when thė wattmeter readings are the same.
 (c) a.c. may be distributed using a single-phase system with 2 wires, a three-phase system with 3 wires or a three-phase system with 4 wires.
 (d) The national standard phase sequence for a three-phase supply is R, Y, B.

Three loads, each of resistance 16 Ω and inductive reactance 12 Ω are connected in delta to a 400 V, 3-phase supply. Determine the quantities stated in problems 7 to 12, selecting the correct answer from the following list.
(a) 4 Ω; (b) $\sqrt{3}$ (400) V; (c) $\sqrt{3}$ (6.4) kW; (d) 20 A; (e) 6.4 kW; (f) $\sqrt{3}$ (20) A; (g) 20 Ω; (h) $\frac{20}{\sqrt{3}}$ A; (i) $\frac{400}{\sqrt{3}}$ V; (j) 19.2 kW; (k) 100 A; (l) 400 V; (m) 28 Ω.

7 Phase impedance.
8 Line voltage.
9 Phase voltage.
10 Phase current.
11 Line current.
12 Total power dissipated in the load.

CONVENTIONAL PROBLEMS

1 Three loads, each of resistance 50 Ω are connected in star to a 400 V, 3-phase supply. Determine (a) the phase voltage; (b) the phase current and (c) the line current. [(a) 231 V; (b) 4.62 A; (c) 4.62 A.]
2 If the loads in *Problem 1* are connected in delta to the same supply determine (a) the phase voltage; (b) the phase current and (c) the line current. [(a) 400 V; (b) 8 A; (c) 13.86 A.]
3 A star-connected load consists of three identical coils, each of inductance

159.2 mH and resistance 50 Ω. If the supply frequency is 50 Hz and the line current is 3 A determine (a) the phase voltage and (b) the line voltage.
[(a) 212 V; (b) 367 V.]

4 Obtain a relationship between the line and phase voltages and line and phase current for a delta connected system. Three inductive loads each of resistance 75 Ω and inductance 318.4 mH are connected in delta to a 415 V, 50 Hz, 3-phase supply. Determine (a) the phase voltage; (b) the phase current, and (c) the line current. [(a) 415 V; (b) 3.32 A; (c) 5.75 A.]

5 Three identical capacitors are connected (a) in star, (b) in delta to a 400 V, 50 Hz 3-phase supply. If the line current is 12 A determine in each case the capacitance of each of the capacitors. [(a) 165.4 μF; (b) 55.13 μF.]

6 Three coils each having resistance 6 Ω and inductance L H are connected (a) in star and (b) in delta, to a 415 V, 50 Hz, 3-phase supply. If the line current is 30 A, find for each connection the value of L.
[(a) 16.78 mH; (b) 73.84 mH]

7 A 400 V, 3 phase, 4 wire, star-connected system supplies three resistive loads of 15 kW, 20 kW and 25 kW in the red, yellow and blue phases respectively. Determine the current flowing in each of the four conductors.
$\begin{bmatrix} I_R = 64.95 \text{ A}, I_Y = 86.60 \text{ A} \\ I_B = 108.25 \text{ A}, I_N = 37.50 \text{ A} \end{bmatrix}$

8 Determine the total power dissipated by three 20 Ω resistors when connected (a) in star and (b) in delta to a 440 V, 3-phase supply.
[(a) 9.68 kW; (b) 29.04 kW]

9 Determine the power dissipated in the circuit of *Problem 3* [1.35 kW]

10 A balanced delta connected load has a line voltage of 400 V, a line current of 8 A and a lagging power factor of 0.94. Draw a complete phasor diagram of the load. What is the total power dissipated by the load? [5.21 kW]

11 A 3-phase, star-connected alternator delivers a line current of 65 A to a balanced delta-connected load at a line voltage of 380 V. Calculate (a) the phase voltage of the alternator, (b) the alternator phase current and (c) the load phase current.
[(a) 219.4 V; (b) 65 A; (c) 37.53 A.]

12 Three inductive loads, each of resistance 4 Ω and reactance 9 Ω are connected in delta. When connected to a 3-phase supply the loads consume 1.2 kW. Calculate (a) the power factor of the load; (b) the phase current; (c) the line current and (d) the supply voltage.
[(a) 0.406; (b) 10 A; (c) 17.32 A; (d) 98.49 V.]

13 The input voltage, current and power to a motor is measured as 415 V, 16.4 A and 6 kW respectively. Determine the power factor of the system. [0.509]

14 A 440 V, 3-phase a.c. motor has a power output of 11.25 kW and operates at a power factor of 0.8 lagging and with an efficiency of 84%. If the motor is delta connected determine (a) the power input; (b) the line current and (c) the phase current. [(a) 13.39 kW; (b) 21.96 A; (c) 12.68 A.]

15 Two wattmeters are connected to measure the input power to a balanced 3-phase load. If the wattmeter readings are 9.3 kW and 5.4 kW determine (a) the total output power; and (b) the load power factor.
[(a) 14.7 kW; (b) 0.909.]

16 8 kW is found by the two-wattmeter method to be the power input to a 3-phase motor. Determine the reading of each wattmeter if the power factor of the system is 0.85. [5.431 kW; 2.569 kW]

17 Show that the power in a three phase balanced system can be measured by two

wattmeters and deduce an expression for the total power in terms of the wattmeter readings. When the two-wattmeter method is used to measure the input power of a balanced load, the readings on the wattmeters are 7.5 kW and 2.5 kW, the connections to one of the coils on the meter reading 2.5 kW having to be reversed. Determine (a) the total input power and (b) the load power factor. [(a) 5 kW; (b) 0.277.]

18 Three similar coils, each having a resistance of 4.0 Ω and an inductive reactance of 3.46 Ω are connected (a) in star and (b) in delta across a 400 V, 3-phase supply. Calculate for each connection the readings on each of two wattmeters connected to measure the power by the two-wattmeter method.
[(a) 17.15 kW, 5.73 kW; (b) 51.46 kW, 17.18 kW.]

19 A 3-phase, star-connected alternator supplies a delta connected load, each phase of which has a resistance of 15 Ω and inductive reactance 20 Ω. If the line voltage is 400 V, calculate (a) the current supplied by the alternator and (b) the output power and kVA rating of the alternator, neglecting any losses in the line between the alternator and the load. [(a) 27.71 A; (b) 11.52 kW; 19.2 kVA]

20 Each phase of a delta connected load comprises a resistance of 40 Ω and a 40 μF capacitor in series. Determine, when connected to a 415 V, 50 Hz, 3-phase supply (a) the phase current; (b) the line current; (c) the total power dissipated and (d) the kVA rating of the load.
[(a) 4.66 A; (b) 8.07 A; (c) 2.605 kW; (d) 5.80 kVA.]

21 (a) State the advantages of three-phase supplies.
(b) Three 24 μF capacitors are connected in star across a 400 V, 50 Hz, 3-phase supply. What value of capacitance must be connected in delta in order to take the same line current? [8 μF]

5 D.C. transients

A. MAIN POINTS CONCERNED WITH D.C. TRANSIENTS

Transients in series connected *C–R* circuits

1 When a d.c. voltage is applied to a capacitor *C*, and resistor *R* connected in series, there is a short period of time immediately after the voltage is connected, during which the current flowing in the circuit and the voltages across *C* and *R* are changing. These changing values are called **transients.**

2 **Charging**

(a) The circuit diagram for a series connected *C–R* circuit is shown in *Fig 1*. When switch S is closed, then by Kirchhoff's voltage law:

$$V = v_C + v_R \qquad (1)$$

(b) The battery voltage V is constant. The capacitor voltage v_C is given by q/C, where q is the charge on the capacitor. The voltage drop across R is given by iR, where i is the current flowing in the circuit. Hence, at all times:

$$V = \frac{q}{C} + iR \qquad (2)$$

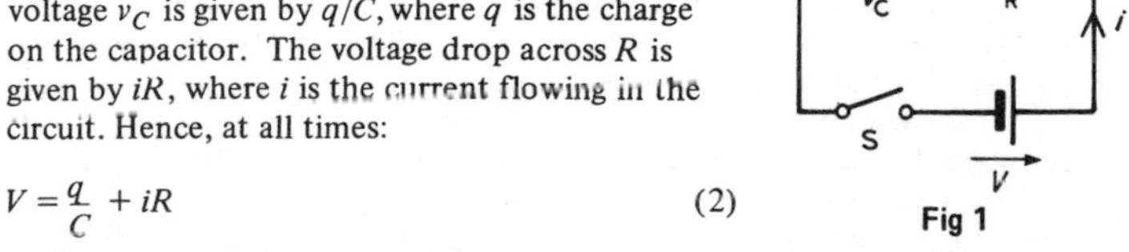

Fig 1

At the instant of closing S, (initial circuit condition), assuming there is no initial charge on the capacitor, q_0 is zero, hence v_{C_0} is zero. Thus from equation (1), $V = 0 + v_{R_0}$, i.e. $v_{R_0} = V$. This shows that the resistance to current is solely due to R, and the initial current flowing, $i_0 = I = V/R$.

(c) A short time later at time t_1 seconds after closing S, the capacitor is partly charged to, say, q_1 coulombs because current has been flowing. The voltage v_{C_1} is now q_1/C volts. If the current flowing is i_1 amperes, then the voltage drop across R has fallen to i_1R volts. Thus, equation (2) is now $V = (q_1/C) + i_1R$.

(d) A short time later still, say at time t_2 seconds after closing the switch, the charge has increased to q_2 coulombs and v_C has increased q_2/C volts. Since $V = v_C + v_R$ and V is a constant, then v_R decreases to i_2R. Thus v_C is increasing and i and v_R are decreasing as time increases.

(e) Ultimately, a few seconds after closing S, (final condition or **steady state** condition), the capacitor is fully charged to, say, Q coulombs, current no longer flows, i.e. $i = 0$, and hence $v_R = iR = 0$. It follows from equation (1) that $v_C = V$.

(f) Curves showing the changes in v_C, v_R and i with time are shown in *Fig 2*.

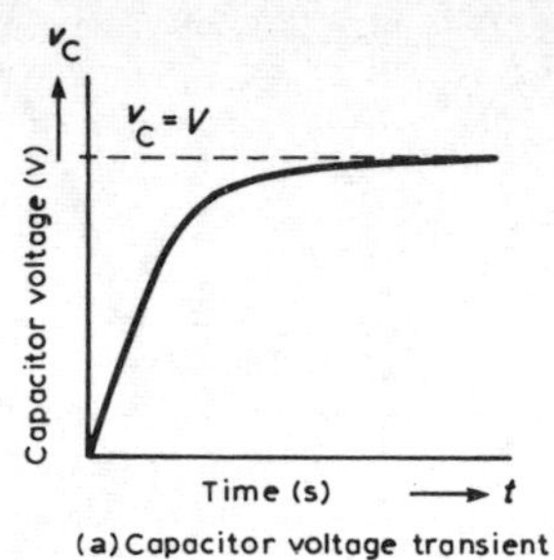

(a) Capacitor voltage transient

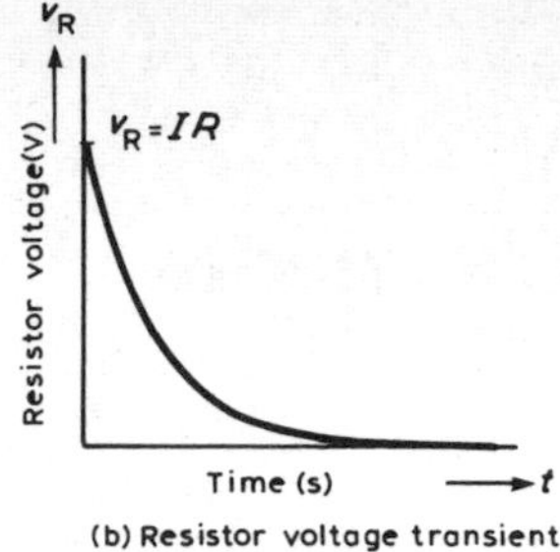

(b) Resistor voltage transient

Fig 2

The curve showing the variation of v_C with time is called an **exponential growth curve** and the graph is called the 'capacitor voltage/time' characteristic. The curves showing the variation of v_R and i with time are called **exponential decay curves**, and the graphs are called 'resistor voltage/time' and 'current/time' characteristics respectively. (The name 'exponential' shows that the shape can be expressed mathematically by an exponential mathematical equation, see para. 5.)

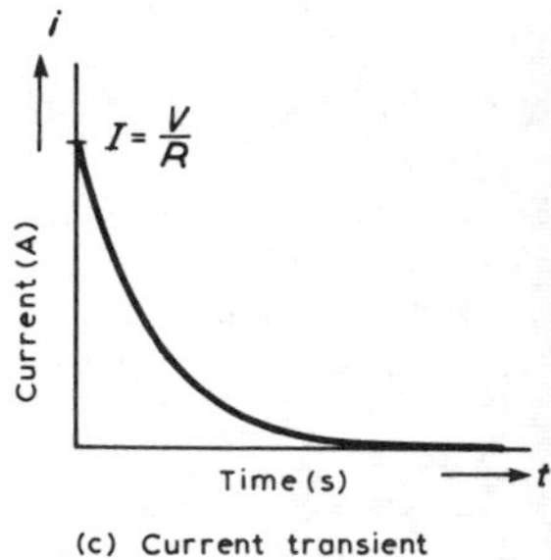

(c) Current transient

3 **The time constant**

(a) If a constant d.c. voltage is applied to a series connected C–R circuit, a transient curve of capacitor voltage v_C is as shown in *Fig 2(a)*.

(b) With reference to *Fig 3*, let the constant voltage supply be replaced by a variable voltage supply at time t_1 seconds. Let the voltage be varied so that the **current** flowing in the circuit is **constant**.

(c) Since the current flowing is a constant, the curve will follow a tangent, AB, drawn to the curve at point A.

(d) Let the capacitor voltage v_C reach its final value of V at time t_2 seconds.

(e) The time corresponding to $(t_2 - t_1)$ seconds is called the **time constant** of the circuit, denoted by the Greek letter 'tau', τ. The value of the time constant is CR seconds, i.e., for a series connected C–R circuit, **time constant,** $\tau = CR$ **seconds.**

Since the variable voltage mentioned in para. 3(b) above can be applied at any instant during the transient change, it may be applied at $t = 0$,

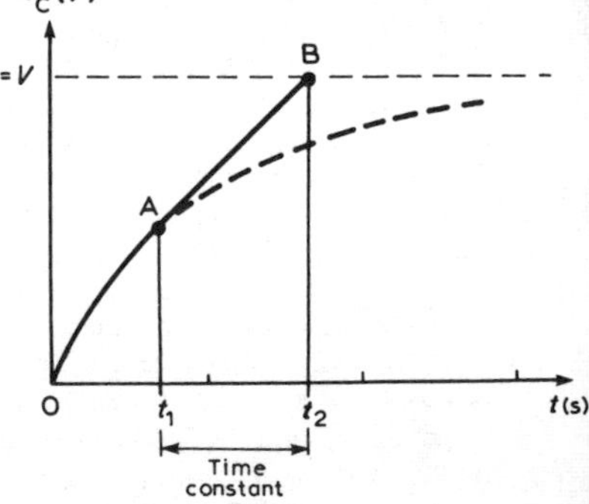

Fig 3

i.e., at the instant of connecting the circuit to the supply. If this is done, then the time constant of the circuit may be defined as:

'the time taken for a transient to reach its final state if the initial rate of change is maintained'.

4 There are two main methods of drawing transient curves graphically.

(a) The tangent method–this method is shown in *Problem 1*.

(b) The initial slope and three point method, this is shown in *Problem 2* and is based on the following properties of a transient exponential curve.

(i) For a growth curve, the value of a transient at a time equal to one time constant is 0.632 of its steady state value (usually taken as 63% of the steady state value); at a time equal to two and a half time constants is 0.918 if its steady state value (usually taken as 92% of its steady state value) and at a time equal to five time constants is equal to its steady state value.

(ii) For a decay curve, the value of a transient at a time equal to one time constant is 0.368 of its initial value (usually taken as 37% of its initial value), at a time equal to two and a half time constants is 0.082 of its initial value (usually taken as 8% of its initial value) and at a time equal to five time constants is equal to zero.

5 The transient curves shown in *Fig 2* have mathematical equations, obtained by solving the differential equations representing the circuit. The equations of the curves are:

growth of capacitor voltage, $v_C = V\left(1-e^{(-\frac{t}{CR})}\right) = V\left(1-e^{(-\frac{t}{\tau})}\right)$

decay of resistor voltage, $v_R = Ve^{(-\frac{t}{CR})} = Ve^{(-\frac{t}{\tau})}$ and

decay of current flowing, $i = Ie^{(-\frac{t}{CR})} = Ie^{(-\frac{t}{\tau})}$

6 **Discharging**

When a capacitor is charged (i.e. with the switch in position A in *Fig 4*), and the switch is then moved to position B, the electrons stored in the capacitor keep the current flowing for a short time. Initially, at the instant of moving from A to B, the current flow is such that the capacitor voltage v_C is balanced by an equal and opposite voltage $v_R = iR$. Since initially $v_C = v_R = V$, then $i = I = V/R$. During the transient decay, by applying Kirchhoff's voltage law to *Fig. 4*, $v_C = v_R$. Finally the transients decay exponentially to zero, i.e. $v_C = v_R = 0$. The transient curves representing the voltages and current are as shown in *Fig 4*.

7 The equations representing the transient curves during the discharge period of a series connected *C–R* circuit are:

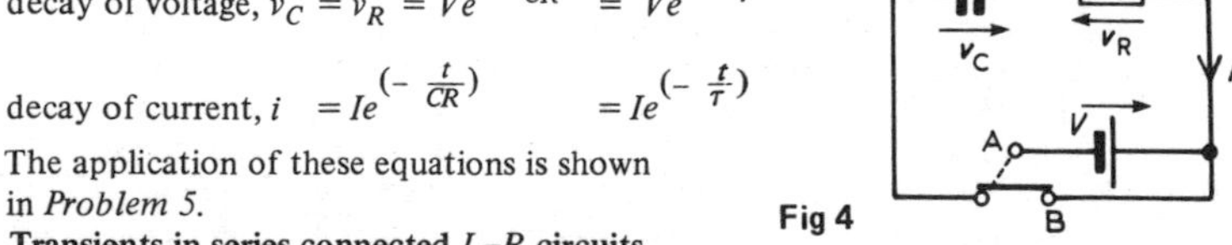

decay of voltage, $v_C = v_R = Ve^{(-\frac{t}{CR})} = Ve^{(-\frac{t}{\tau})}$

decay of current, $i = Ie^{(-\frac{t}{CR})} = Ie^{(-\frac{t}{\tau})}$

The application of these equations is shown in *Problem 5*.

Fig 4

Transients in series connected *L–R* circuits

8 When a d.c. voltage is connected to a circuit having inductance L connected in series with resistance R, there is a short period of time immediately after the

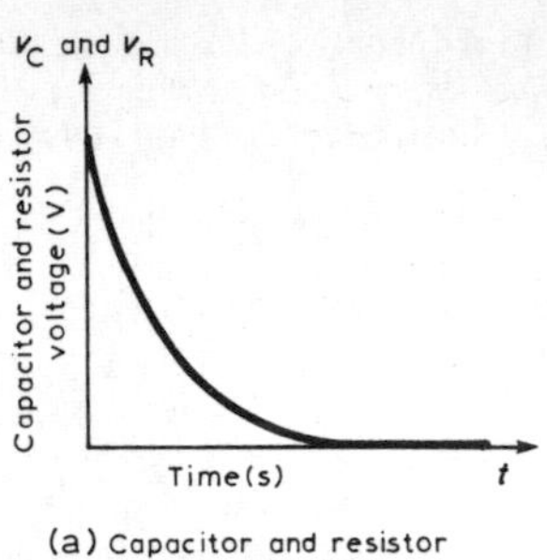

(a) Capacitor and resistor voltage transient

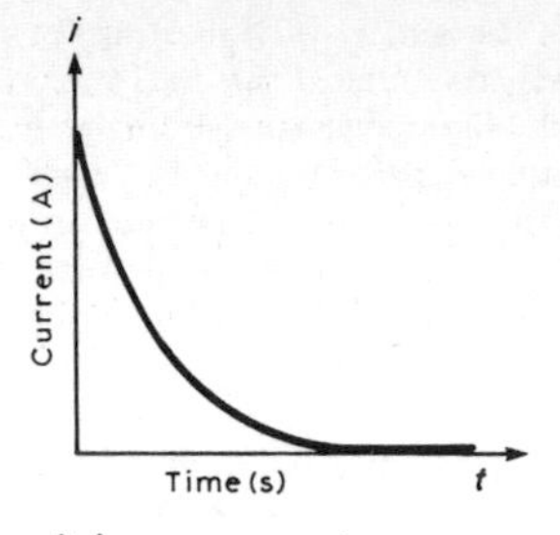

(b) Current transient

Fig 5

voltage is connected, during which the current flowing in the circuit and the voltages across L and R are changing. These changing values are called **transients.**

9 **Current growth**

(a) The circuit diagram for a series connected L–R circuit is shown in *Fig 6*. When switch S is closed, then by Kirchhoff's voltage law:
$V = v_L + v_R$ (3)

(b) The battery voltage V is constant. The voltage of the inductance is the induced voltage, i.e.

$$v_L = L \times \frac{\text{change of current}}{\text{change of time}}$$

shown as $L(di/dt)$. The voltage drop across R, v_R is given by iR. Hence, at all times:
$V = L(di/dt) + iR$ (4)

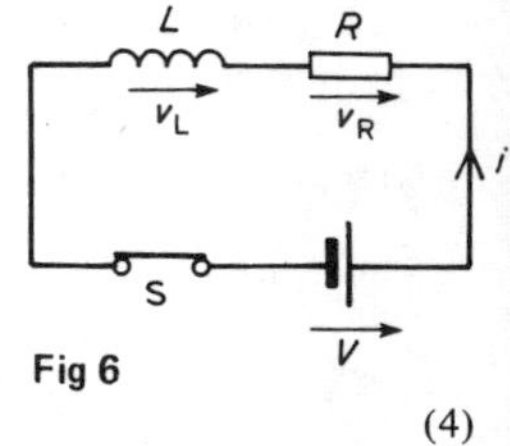

Fig 6

(c) At the instant of closing the switch, the rate of change of current is such that it induces an e.m.f. in the inductance which is equal and opposite to V, hence $V = v_L + 0$, i.e. $v_L = V$. From equation (3), because $v_L = V$, then $v_R = 0$ and $i = 0$.

(d) A short time later at time t_1 seconds after closing S, current i_1 is flowing, since there is a rate of change of current initially, resulting in a voltage drop of $i_1 R$ across the resistor. Since V (constant) $= v_L + v_R$ the induced e.m.f. is reduced and equation (4) becomes:

$$V = L\frac{di_1}{dt_1} + i_1 R.$$

(e) A short time later still, say at time t_2 seconds after closing the switch, the current flowing is i_2, and the voltage drop across the resistor increases to $i_2 R$. Since v_R increases, v_L decreases.

(f) Ultimately, a few seconds after closing S, the current flow is entirely limited by R, the rate of change of current is zero and hence v_L is zero. Thus $V = iR$. Under these conditions, steady state current flows, usually signified by I. Thus, $I = V/R$, $v_R = IR$ and $v_L = 0$ at steady state conditions.

(g) Curves showing the changes in v_L, v_R and i with time are shown in *Fig 7* and indicate that v_L is a maximum value initially (i.e. equal to V), decaying exponentially to zero, whereas v_R and i grow exponentially from zero to their steady state values of V and $I = V/R$ respectively.

10 **The time constant**

With reference to para. 3, the time constant of a series connected L–R circuit is defined in the same way as the time constant for a series connected C–R circuit. Its value is given by:

time constant, $\tau = L/R$ seconds.

11 Transient curves representing the induced voltage/time, resistor voltage/time and current/time characteristics may be drawn graphically, as outlined in para. 4. The methods of construction are shown in detail in *Problems 6 and 7.*

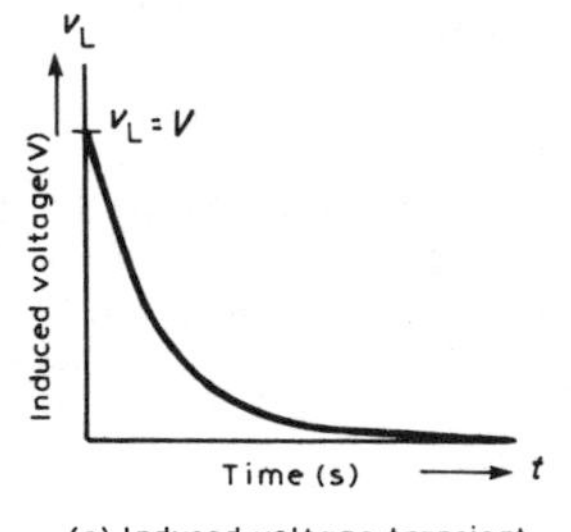

(a) Induced voltage transient

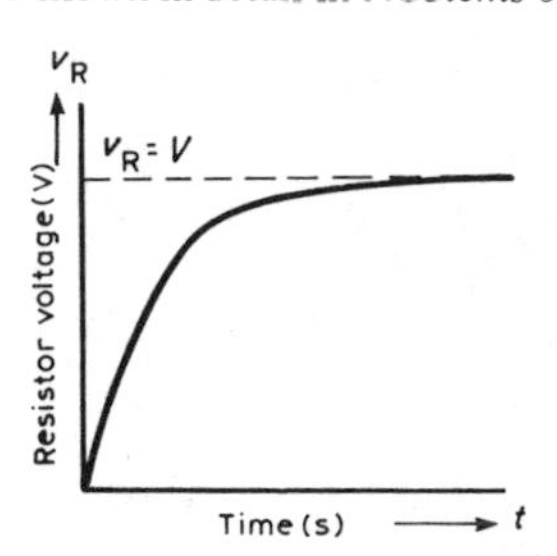

(b) Resistor voltage transient

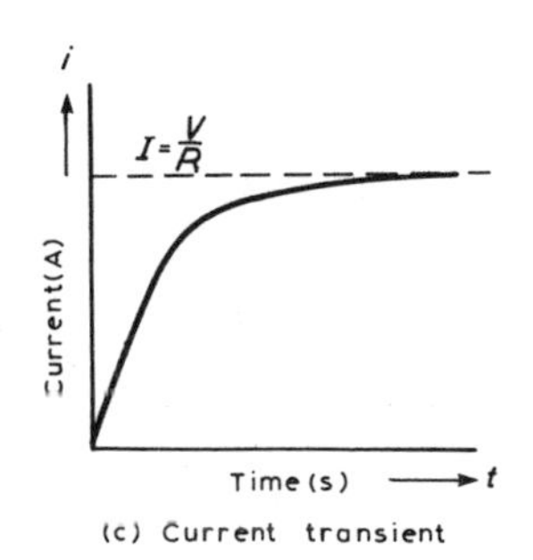

(c) Current transient

Fig 7

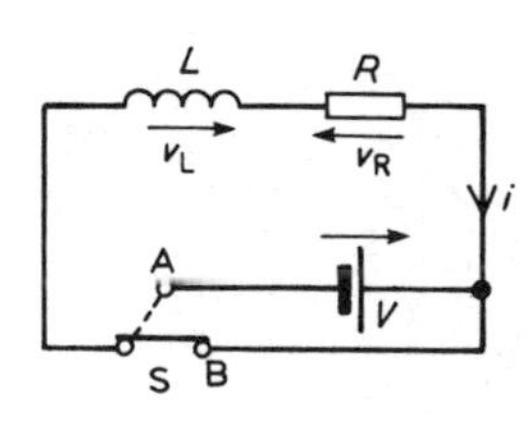

Fig 8

12 Each of the transient curves shown in *Fig 7* have mathematical equations, and these are:

decay of induced voltage, $v_L = Ve^{(-\frac{Rt}{L})} = Ve^{(-\frac{t}{\tau})}$,

growth of resistor voltage, $v_R = V(1-e^{-\frac{Rt}{L}}) = V(1-e^{-\frac{t}{\tau}})$

growth of current flow, $i = I(1-e^{-\frac{Rt}{L}}) = I(1-e^{-\frac{t}{\tau}})$

The application of these equations is shown in *Problems 8 to 10.*

13 **Current decay**

When a series connected L–R circuit is connected to a d.c. supply as shown with S in position A of *Fig 8*, a current $I = V/R$ flows after a short time, creating a magnetic field ($\Phi \propto I$) associated with the inductor. When S is moved to position B, the current value decreases, causing a decrease in the strength of the magnetic field. Flux linkages occur, generating a voltage v_L, equal to $L(di/dt)$. By Lenz's law, this voltage keeps current i flowing in the circuit, its value being limited by R. Thus $v_L = v_R$. The current decays exponentially to zero and since v_R is proportional to the current flowing, v_R decays exponentially to zero. Since $v_L = v_R$, v_L also decays exponentially to zero. The curves representing these transients are similar to those shown in *Fig 5*.

14 The equations representing the decay transient curves are:

decay of voltages, $v_L = v_R = Ve^{\left(-\frac{Rt}{L}\right)} = Ve^{\left(-\frac{t}{\tau}\right)}$

decay of current, $i = Ie^{\left(-\frac{Rt}{L}\right)} = Ie^{\left(-\frac{t}{\tau}\right)}$

The application of these equations is also shown in *Problems 8 to 10*.

Fig 9

The effects of time constant on a rectangular wave

15 By varying the value of either C or R in a series connected C–R circuit, the time constant ($\tau = CR$), of a circuit can be varied. If a rectangular waveform varying from $+E$ to $-E$ is applied to a C–R circuit as shown in *Fig 9*, output waveforms of the capacitor voltage have various shapes, depending on the value of R. When R is small, $\tau = CR$ is small and an output waveform such as that shown in *Fig 10(a)* is obtained. As the value of R is increased, the waveform changes to that shown in *Fig 10(b)*. When R is large, the waveform is as shown in *Fig 10(c)*, the circuit then being described as an **integrator circuit**.

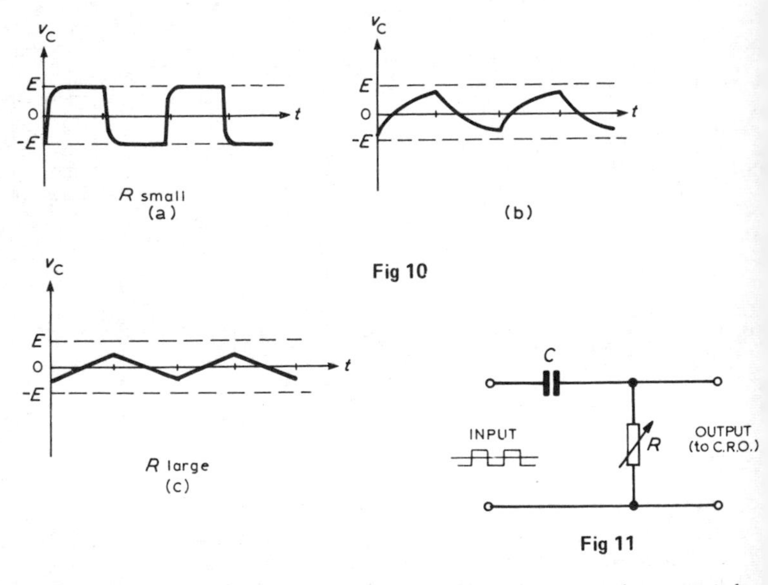

Fig 10

Fig 11

16 If a rectangular waveform varying from $+E$ to $-E$ is applied to a series connected C–R circuit and the waveform of the voltage drop across the resistor is observed, as shown in *Fig 11*, the output waveform alters as R is varied due to the time constant, ($\tau = CR$), altering. When R is small, the waveform is as shown in *Fig 12(a)*, the voltage being generated across R by the capacitor discharging fairly quickly. Since the change in capacitor voltage is from $+E$ to $-E$, the

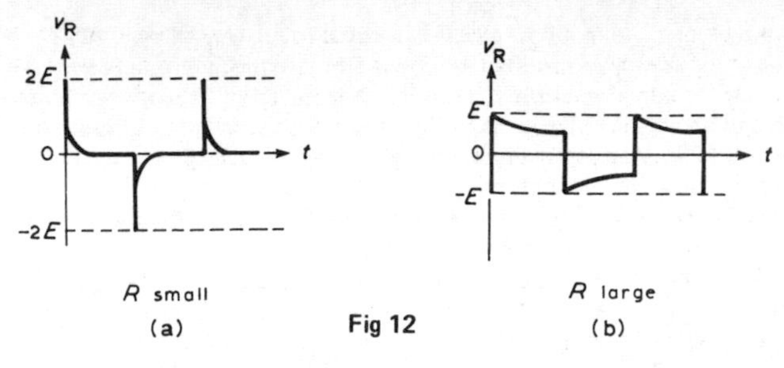

Fig 12

change in discharge current is $2E/R$, resulting in a change in voltage across the resistor of $2E$. This circuit is called a **differentiator circuit**. When R is large, the waveform is as shown in *Fig 12(b)*.

B. WORKED PROBLEMS ON D.C. TRANSIENTS

TRANSIENTS IN SERIES CONNECTED *C–R* CIRCUITS

Problem 1 **A 15 μF uncharged capacitor is connected in series with a 47 kΩ resistor across a 120 V, d.c. supply. Use the tangential graphical method to draw the capacitor voltage/time characteristic of the circuit. From the characteristic, determine the capacitor voltage for a time equal to one time constant after being connected to the supply and also two seconds after being connected to the supply. Also find the time for the capacitor voltage to reach one half of its steady state value.**

To construct an exponential growth curve, the time constant of the circuit and steady state value need to be determined.

Time constant $= CR = 15\ \mu\text{F} \times 47\ \text{k}\Omega = 15 \times 10^{-6} \times 47 \times 10^{3}$
$= 0.705$ s.

Steady state value of v_C is $v_C = V$, i.e. $v_C = 120$ V

With reference to *Fig 13*, the scale of the horizontal axis is drawn so that it spans at least five time constants, i.e. 5×0.705 or about $3\frac{1}{2}$ seconds. The scale of the vertical axis spans the change in the capacitor voltage, that is, from 0 to 120 V.

A broken line AB is drawn corresponding to the final value of v_C.

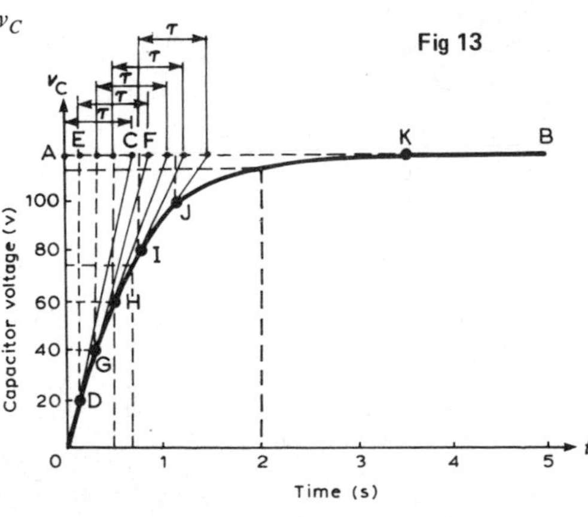

Fig 13

Point C is measured along AB so that AC is equal to 1τ, i.e., AC = 0.705 s. Straight line OC is drawn. Assuming that about five intermediate points are needed to draw the curve accurately, a point D is selected on OC corresponding to a v_C value of about 20 V. DE is drawn vertically. EF is made to correspond to 1τ, i.e., EF = 0.705 s. A straight line is drawn joining DF. This procedure of
(a) drawing a vertical line through point selected,
(b) at the steady-state value, drawing a horizontal line corresponding to 1τ, and
(c) joining the first and last points,
is repeated for v_C values of 40, 60, 80 and 100 V, giving points G, H, I and J.

The capacitor voltage effectively reaches its steady-state value of 120 V after a time equal to five time constants, shown as point K. Drawing a smooth curve through points O, D, G, H, I, J and K gives the exponential growth curve of capacitor voltage.

From the graph, the value of capacitor voltage at a time equal to the time constant is about **75 V**. It is a characteristic of all exponential growth curves, that after a time equal to one time constant, the value of the transient is 0.632 of its steady-state value. In this problem, $0.632 \times 120 = 75.84$ V. Also from the graph, when t is two seconds, v_C is about **115 Volts.** [This value may be checked using the equation $v_C = V(1-e^{-t/\tau})$, where $V = 120$ V, $\tau = 0.705$ s and $t = 2$ s. This calculation gives $v_C = 112.97$ V.]

The time for v_C to rise to one half of its final value, i.e., 60 V, can be determined from the graph and is about **0.5 s**. [This value may be checked using $v_C = V(1-e^{-t/\tau})$ where $V = 120$ V, $v_C = 60$ V and $\tau = 0.705$ s, giving $t = 0.489$ s.]

Problem 2 A 4 μF capacitor is charged to 24 V and then discharged through a 220 kΩ resistor. Use the 'initial slope and three point' method to draw: (a) the capacitor voltage/time characteristic; (b) the resistor voltage/time characteristic and (c) the current/time characteristic, for the transients which occur. From the characteristics determine the values of capacitor voltage, resistor voltage and current one and a half seconds after discharge has started.

To draw the transient curves, the time constant of the circuit and steady state values are needed.

Time constant, $\tau = CR = 4 \times 10^{-6} \times 220 \times 10^3 = 0.88$ s.

Initially, capacitor voltage $v_C = v_R = 24$ V. $i = \dfrac{V}{R} = \dfrac{24}{220 \times 10^3} = 0.109$ mA

Finally, $v_C = v_R = i = 0$.

(a) The exponential decay of capacitor voltage is from 24 V to 0 V in a time equal to five time constants, i.e., $5 \times 0.88 = 4.4$ s. With reference to *Fig 14*, to construct the decay curve:
   (i) the horizontal scale is made so that it spans at least five time constants, i.e. 4.4 s,
   (ii) the vertical scale is made to span the change in capacitor voltage, i.e., 0 to 24 V,
   (iii) point A corresponds to the initial capacitor voltage, i.e., 24 V,
   (iv) OB is made equal to one time constant and line AB is drawn. This gives the initial slope of the transient.
   (v) the value of the transient after a time equal to one time constant is 0.368 of the initial value (see para. 4), i.e., $0.368 \times 24 = 8.83$ V.

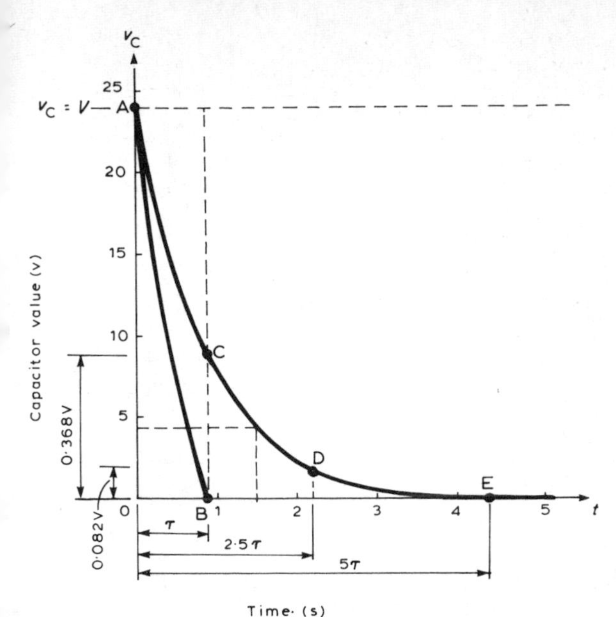

Fig 14

A vertical line is drawn through B and distance BC is made equal to 8.83 V,

(vi) the value of the transient after a time equal to two and a half time constants is 0.082 of the initial value, i.e., $0.082 \times 24 = 1.97$ V, shown as point D in *Fig 14*,

(vii) the transient effectively dies away to zero after a time equal to five time constants, i.e., 4.4 s, giving point E.

The smooth curve drawn through points A, C, D and E represents the decay transient. At $1\frac{1}{2}$ s after decay has started, $v_C \triangleq$ **4.4 V**. [This may be checked using $v_C = Ve^{-t/\tau}$, where $V = 24$, $t = 1\frac{1}{2}$ and $\tau = 0.88$, giving $v_C = 4.36$ V.]

(b) The voltage drop across the resistor is equal to the capacitor voltage when a capacitor is discharging through a resistor, thus the resistor voltage/time characteristic is identical to that shown in *Fig 14*. Since $v_R = v_C$, then at $1\frac{1}{2}$ seconds after decay has started, $v_R \triangleq$ **4.4 V** (see (a) above).

(c) The current/time characteristic is constructed in the same way as the capacitor voltage/time characteristic, shown in part (a) of this problem, and is as shown in *Fig 15*. The values are:
point A: initial value of current = 0.109 mA
point C: at 1τ, i = $0.368 \times 0.109 = 0.040$ mA
point D: at 2.5τ, i = $0.082 \times 0.109 = 0.009$ mA
point E: at 5τ, i = 0.
Hence current transient is as shown. At a time of $1\frac{1}{2}$ seconds, the value of current, from the characteristic is **0.02 mA**. [This may be checked using $i = Ie^{(-t/\tau)}$ where $I = 0.109$, $t = 1\frac{1}{2}$ and $\tau = 0.88$, giving $i = 0.0198$ mA or 19.8 μA.]

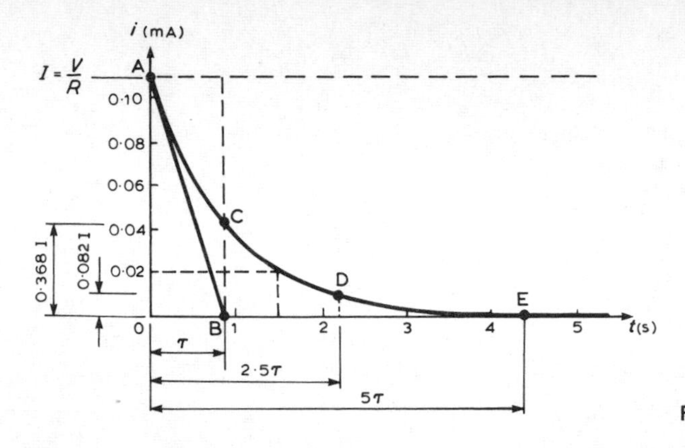

Fig 15

Problem 3 A 20 μF capacitor is connected in series with a 50 kΩ resistor and the circuit is connected to a 20 V, d.c. supply. Determine
(a) the initial value of the current flowing;
(b) the time constant of the circuit;
(c) the value of the current one second after connection;
(d) the value of the capacitor voltage two seconds after connection and
(e) the time after connection when the resistor voltage is 15 V.

Parts (c), (d) and (e) may be determined graphically, as shown in *Problems 1 and 2* or by calculation as shown below.

$V = 20$ V, $C = 20\ \mu$F $= 20 \times 10^{-6}$ F, $R = 50$ kΩ $= 50 \times 10^3\ \Omega$

(a) With reference to para. 2(b), the initial value of the current flowing is

$$I = \frac{V}{R}, \text{ i.e. } \frac{20}{50 \times 10^3} = \mathbf{0.4\ mA}$$

(b) From para. 3(e) the time constant, $\tau = CR = (20 \times 10^{-6}) \times (50 \times 10^3)$
$= \mathbf{1\ s}$

(c) From para. 5, $i = Ie^{-t/\tau}$.
Working in mA units, $i = 0.4e^{-1/1} = 0.4 \times 0.368 = \mathbf{0.147\ mA}$

(d) From para. 5, $v_C = V(1-e^{-t/\tau}) = 20(1-e^{-2/1})$
$= 20(1-0.135) = 20 \times 0.865$
$= \mathbf{17.3\ V}$

(e) From para. 5, $v_R = Ve^{-t/\tau}$
Thus $15 = 20e^{-t/1}$, $\frac{15}{20} = e^{-t}$, i.e. $e^t = \frac{20}{15} = \frac{4}{3}$.

Taking natural logarithms of each side of the equation gives $t = \ln \frac{4}{3} = \ln 1.3333$
$t = \mathbf{0.288\ s}$

Problem 4 A circuit consists of a resistor connected in series with a 0.5 μF capacitor and has a time constant of 12 ms. Determine (a) the value of the resistor and (b) the capacitor voltage 7 ms after connecting the circuit to a 10 V supply.

(a) The time constant $\tau = CR$, hence $R = \frac{\tau}{C}$

i.e. $R = \frac{12 \times 10^{-3}}{0.5 \times 10^{-6}} = 24 \times 10^3 = \mathbf{24\ k\Omega}$

(b) The equation for the growth of capacitor voltage is:

$v_C = V(1 - e^{-t/\tau})$

Since $\tau = 12$ ms $= 12 \times 10^{-3}$ s, $V = 10$ V and $t = 7$ ms $= 7 \times 10^{-3}$ s,

then $v_C = 10\left(1 - e^{-\frac{7 \times 10^{-3}}{12 \times 10^{-3}}}\right) = 10(1 - e^{-0.58\dot{3}})$

$= 10(1 - 0.558) \qquad = \mathbf{4.42\ V}$

[The value of $e^{-0.58\dot{3}}$ can be determined either by using a calculator or as follows: Let $y = e^{-0.58\dot{3}}$

$\ln y = \ln e^{-0.58\dot{3}}$

i.e. $\ln y = -0.58\dot{3}$, by the rules of logarithms.

Using natural log tables, the antilog of -0.5833, i.e. $\bar{1}.4167$ is 0.5580.

Thus $v_C = 10(1 - 0.5580)$

$= \mathbf{4.42\ V}$]

Alternatively, the value of v_C when t is 7 ms may be determined using the growth characteristic as shown in *Problem 1.*

Problem 5 A capacitor is charged to 100 V and then discharged through a 50 kΩ resistor. If the time constant of the circuit is 0.8 s, determine:
(a) the value of the capacitor;
(b) the time for the capacitor voltage to fall to 20 V;
(c) the current flowing when the capacitor has been discharging for 0.5 s and
(d) the voltage drop across the resistor when the capacitor has been discharging for one second.

Parts (b), (c) and (d) of this problem may be solved graphically as shown in *Problems 1 and 2* or by calculation as shown below.

$V = 100$ V, $\tau = 0.8$ s, $R = 50$ kΩ $= 50 \times 10^3$ Ω

(a) Since time constant, $\tau = CR$, $C = \tau/R$

$$\text{i.e. } C = \frac{0.8}{50 \times 10^3} = \mathbf{16\ \mu F}$$

(b) $v_C = Ve^{-t/\tau}$

$20 = 100e^{-t/0.8}$, i.e. $\frac{1}{5} = e^{-t/0.8}$

Thus $e^{t/0.8} = 5$ and taking natural logs of each side, gives

$\frac{t}{0.8} = \ln 5$, i.e., $t = 0.8 \ln 5$

Hence $t = \mathbf{1.29\ s}$

(c) $i = Ie^{-t/\tau}$

The initial current flowing, $I = \frac{V}{R} = \frac{100}{50 \times 10^3} = 2$ mA

Working in mA units, $i = Ie^{-t/\tau} = 2e^{\left(-\frac{0.5}{0.8}\right)} = 2e^{-0.625}$

$= 2 \times 0.535 = \mathbf{1.07\ mA}$

(d) $v_R = v_C = Ve^{-t/\tau}$

$$= 100e^{-\frac{1}{0.8}} = 100e^{-1.25}$$
$$= 100 \times 0.287 = 28.7 \text{ V}$$

TRANSIENTS IN SERIES CONNECTED L–R CIRCUITS

Problem 6 A relay has an inductance of 100 mH and a resistance of 20 Ω. It is connected to a 60 V, d.c. supply. Use the 'initial slope and three point' method to draw the current/time characteristic and hence determine the value of current flowing at a time equal to two time constants and the time for the current to grow to 1.5 A.

Before the current/time characteristic can be drawn; the time constant and steady-state value of the current have to be calculated.

Time constant, $\tau = \frac{L}{R} = \frac{100 \times 10^{-3}}{20} = 5 \text{ ms}$

Final value of current, $I = \frac{V}{R} = \frac{60}{20} = 3 \text{ A}$

The method used to construct the characteristic is the same as that used in *Problem 2.*

(a) The scales should span at least five time constants (horizontally), i.e. 25 ms, and 3 A (vertically).

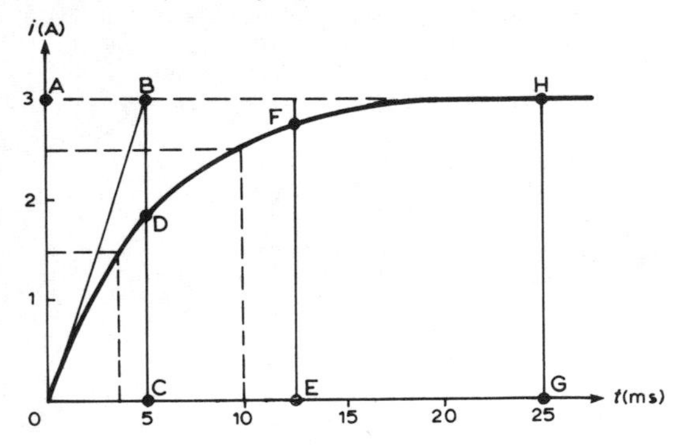

Fig 16

(b) With reference to *Fig 16*, the initial slope is obtained by making AB equal to 1 time constant, (5 ms), and joining OB.

(c) At a time of 1 time constant, CD is $0.632 \times I = 0.632 \times 3 = 1.896$ A
At a time of 2.5 time constants, EF is $0.918 \times I = 0.918 \times 3 = 2.754$ A
At a time of 5 time constants, GH is $I = 3$ A

(d) A smooth curve is drawn through points 0, D, F and H and this curve is the current/time characteristic.

From the characteristic, when $t = 2\tau$, $i \simeq$ **2.6 A**. [This may be checked by calculation using $i = I(1-e^{-t/\tau})$, where $I = 3$ and $t = 2\tau$, giving $i = 2.59$ A]. Also, when the current is 1.5 A, the corresponding time is about **3.6 ms.** [This may be checked by calculation, using $i = I(1-e^{-t/\tau})$ where $i = 1.5, I = 3$ and $\tau = 5$ ms, giving $t = 3.466$ ms.]

Problem 7 The field winding of a 110 V, d.c. motor has a resistance of 15 Ω and a time constant of 2 s. Determine the inductance and use the tangential method to draw the current/time characteristic when the supply is removed and replaced by a shorting link. From the characteristic determine (a) the current flowing in the winding 3 s after being shorted-out and (b) the time for the current to decay to 5 A.

Since the time constant, $\tau = \frac{L}{R}$, $L = R\tau$

i.e. inductance, $L = 15 \times 2 = $ **30 H.**

The current/time characteristic is constructed in a similar way to that used in *Problem 1.*

(i) The scales should span at least five time constants horizontally, i.e. 10 s, and $I = V/R = 110/15 = 7.\dot{3}$ A vertically.

(ii) With reference to *Fig 17*, the initial slope is obtained by making OB equal to 1 time constant, (2 s), and joining AB.

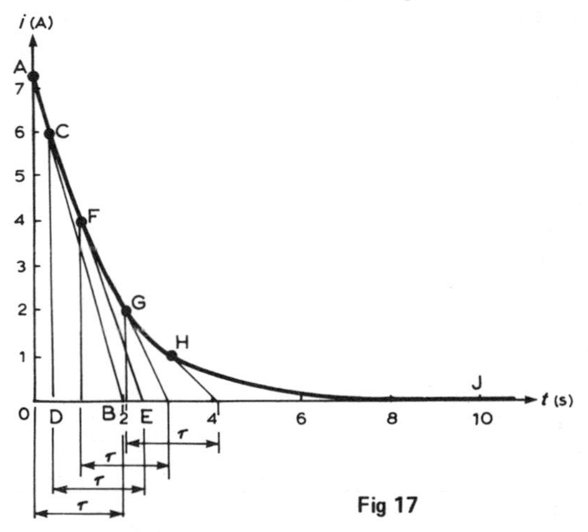

Fig 17

(iii) At, say, $i = 6$ A, let C be the point on AB corresponding to a current of 6 A. Make DE equal to 1 time constant, (2 s), and join CE.

(iv) Repeat the procedure given in (iii) for current values of, say, 4 A, 2 A and 1 A, giving points F, G and H.

(v) Point J is at five time constants, when the value of current is zero.

(vi) Join points A, C, F, G, H and J with a smooth curve, This curve is the current/time characteristic.

(a) From the current/time characteristic, when $t = 3$ s, $i =$ **1.2 A.** [This may be checked by calculation using $i = Ie^{-t/\tau}$, where $I = 7.\dot{3}$, $t = 3$ and $\tau = 2$, giving $i = 1.64$ A.] The discrepancy between the two results is due to relatively few values, such as C, F, G and H, being taken.

(b) From the characteristic, when $i = 5$ A, $t =$ **0.65 s**. [This may be checked by calculation using $i = Ie^{-t/\tau}$, where $i = 5, I = 7.\dot{3}$ and $\tau = 2$, giving $t = 0.766$ s.] The discrepancy between the graphical and calculated values is due to relatively few values such as C, F, G and H being taken.

Problem 8 The winding of an electromagnet has an inductance of 3 H and a resistance of 15 Ω. When it is connected to a 120 V, d.c. supply, calculate:
(a) the steady state value of current flowing in the winding;
(b) the time constant of the circuit;
(c) the value of the induced e.m.f. after 0.1 s;
(d) the time for the current to rise to 85% of its final value; and
(e) the value of the current after 0.3 s.

(a) The steady state value of current is $I = V/R$, i.e. $I = 120/15 =$ **8 A**
(b) The time constant of the circuit, $\tau = L/R = 3/15 =$ **0.2 s**
Parts (c), (d) and (e) of this problem may be determined by drawing the transients graphically as shown in *Problems 6 and 7* or by calculation as shown below.
(c) The induced e.m.f., v_L is given by $v_L = Ve^{-t/\tau}$. The d.c. voltage V is 120 V, t is 0.1 s and τ is 0.2 s, hence

$$v_L = 120e^{-0.1/0.2} = 120e^{-0.5} = 120 \times 0.6065$$

i.e. $v_L =$ **72.78 V**
(d) When the current is 85% of its final value, $i = 0.85\ I$.
Also $i = I(1-e^{-t/\tau})$, thus $0.85I = I(1-e^{-t/\tau})$

$$0.85 = 1-e^{-t/\tau} \text{ and } \tau = 0.2, \text{ hence}$$

$$0.85 = 1-e^{-t/0.2}$$

$$e^{-t/0.2} = 1-0.85 = 0.15$$

$$e^{t/0.2} = \frac{1}{0.15} = 6.\dot{6}$$

Taking natural logarithms of each side of this equation gives:

$$\ln e^{t/0.2} = \ln 6.\dot{6}, \text{ and by the rules of logarithms}$$

$$\frac{t}{0.2} \ln e = \ln 6.\dot{6}. \text{ But } \ln e = 1, \text{ hence}$$

$$t = 0.2 \ln 6.\dot{6} \text{ i.e. } t = \mathbf{0.379\ s}$$

(e) The current at any instant is given by $i = I(1-e^{-t/\tau})$
When $I = 8$, $t = 0.3$ and $\tau = 0.2$, then

$$i = 8(1-e^{-\frac{0.3}{0.2}}) = 8(1-e^{-1.5})$$

$$= 8(1-0.2231) = 8 \times 0.7769 \text{ i.e. } i = \mathbf{6.215\ A}$$

Problem 9 A coil having an inductance of 6 H and a resistance of R Ω is connected in series with a resistor of 10 Ω to a 120 V, d.c. supply. The time constant of the circuit is 300 ms. When steady-state conditions have been reached, the supply is replaced instantaneously by a short-circuit. Determine:
(a) the resistance of the coil;
(b) the current flowing in the circuit one second after the shorting link has been placed in the circuit; and
(c) the time taken for the current to fall to 10% of its initial value.

(a) The time constant $\tau = \dfrac{\text{circuit inductance}}{\text{total circuit resistance}} = \dfrac{L}{R+10}$

Thus $R = \dfrac{L}{\tau} - 10 = \dfrac{6}{0.3} - 10 = \mathbf{10\ \Omega}$

Parts (b) and (c) may be determined graphically as shown in *Problems 6 and 7* or by calculation as shown below.

(b) The steady-state current, $I = \dfrac{V}{R} = \dfrac{120}{10+10} = 6$ A

The transient current after 1 second, $i = Ie^{-t/\tau} = 6e^{-1/0.3}$

Thus $i = 6e^{-3.\dot{3}} = 6 \times 0.03567$
$= \mathbf{0.214\ A}$

(c) 10% of the initial value of the current is $10/100 \times 6$, i.e. 0.6 A
Using the equation $i = Ie^{-t/\tau}$, gives
$0.6 = 6e^{-t/0.3}$,

i.e. $\dfrac{0.6}{6} = e^{-t/0.3}$ or $e^{t/0.3} = \dfrac{6}{0.6} = 10$

Taking natural logarithms of each side of this equation gives:
$\dfrac{t}{0.3} = \ln 10$

$t = 0.3 \ln 10 = \mathbf{0.691\ s}$

Problem 10 An inductor has a negligible resistance and an inductance of 200 mH and is connected in series with a 1 kΩ resistor to a 24 V, d.c. supply. Determine the time constant of the circuit and the steady-state value of the current flowing in the circuit. Find (a) the current flowing in the circuit at a time equal to one time constant, (b) the voltage drop across the inductor at a time equal to two time constants and (c) the voltage drop across the resistor after a time equal to three time constants.

The time constant, $\tau = \dfrac{L}{R} = \dfrac{0.2}{1000} = \mathbf{0.2\ ms}$

The steady-state current $I = \dfrac{V}{R} = \dfrac{24}{1000} = \mathbf{24\ mA}$

(a) The transient current, $i = I(1-e^{-t/\tau})$ and $t = 1\tau$
Working in mA units gives, $i = 24(1-e^{-(1\tau/\tau)}) = 24(1-e^{-1})$
$= 24(1-0.368) = \mathbf{15.17\ mA}$

(b) The voltage drop across the inductor, $v_L = Ve^{-t/\tau}$
$= 24e^{-2\tau/\tau} = 24e^{-2}$
$= \mathbf{3.248\ V}$

(c) The voltage drop across the resistor, $v_R = V(1-e^{-t/\tau})$
$= 24(1-e^{-3\tau/\tau}) = 24(1-e^{-3})$
$= \mathbf{22.81\ V}$

C. FURTHER PROBLEMS

SHORT ANSWER PROBLEMS

A capacitor of capacitance C farads is connected in series with a resistor of R ohms and is switched across a constant voltage d.c. supply of V volts. After a time of t seconds, the current flowing is i amperes. Use this data to answer *Problems 1 to 10.*

1 The voltage drop across the resistor at time t seconds is v_R = ...
2 The capacitor voltage at time t seconds is v_C = ...
3 The voltage equation for the circuit is V = ...
4 The time constant for the circuit is τ = ...
5 The final value of the current flowing is . . . .
6 The initial value of the current flowing is I = ...
7 The final value of capacitor voltage is . . . .
8 The initial value of capacitor voltage is . . . .
9 The final value of the voltage drop across the resistor is . . . .
10 The initial value of the voltage drop across the resistor is . . . .

A capacitor charged to V volts is disconnected from the supply and discharged through a resistor of R ohms. Use this data to answer *Problems 11 to 15*

11 The initial value of current flowing is I = ...
12 The approximate time for the current to fall to zero in terms of C and R is . . . . seconds.
13 If the value of resistance R is doubled, the time for the current to fall to zero is . . . . when compared with the time in *Problem 12* above.
14 The approximate fall in the value of the capacitor voltage in a time equal to one time constant is . . . . %.
15 The time constant of the circuit is given by . . . . seconds.

An inductor of inductance L henrys and negligible resistance is connected in series with a resistor of resistance R ohms and is switched across a constant voltage d.c. supply of V volts. After a time interval of t seconds, the transient current flowing is i amperes. Use this data to answer *Problems 16 to 25.*

16 The induced e.m.f., v_L opposing the current flow when $t = 0$ is . . . .
17 The voltage drop across the resistor when $t = 0$ is v_R = ...
18 The current flowing when $t = 0$ is . . . .
19 V, v_R and v_L are related by the equation V = ...
20 The time constant of the circuit in terms of L and R is . . . .
21 The steady-state value of the current is reached in practice in a time equal to . . . . seconds.
22 The steady-state voltage across the inductor is . . . . volts.
23 The final value of the current flowing is . . . . amperes.
24 The steady-state resistor voltage is . . . . volts.
25 The e.m.f. induced in the inductor during the transient in terms of current, time and inductance is . . . . volts.

A series connected L–R circuit carrying a current of I amperes is suddenly short-circuited to allow the current to decay exponentially. Use this data to answer *Problems 26 to 30.*

26 The current will fall to . . . . % of its initial value in a time equal to the time constant.
27 The voltage equation of the circuit is . . . . = ...
28 The time constant of the circuit in terms of L and R is . . . .
29 The current reaches zero in a time equal to . . . . seconds.
30 If the value of R is halved, the time for the current to fall to zero is . . . . when compared with the time in *Problem 29.*

MULTI-CHOICE PROBLEMS (answers on page 191)

An uncharged 2 μF capacitor is connected in series with a 5 MΩ resistor to a 100 V, constant voltage, d.c. supply. In *Problems 1 to 7*, use this data to select the correct answer from those given below.

(a) 10 ms; (b) 100 V; (c) 10 s; (d) 10 V; (e) 20 μA; (f) 1 s; (g) 0 V;
(h) 50 V; (i) 1 ms; (j) 50 μA; (k) 20 mA; (l) 0 A.

1 Determine the time constant of the circuit.
2 Determine the final voltage across the capacitor.
3 Determine the initial voltage across the resistor.
4 Determine the final voltage across the resistor.
5 Determine the initial voltage across the capacitor.
6 Determine the initial current flowing in the circuit.
7 Determine the final current flowing in the circuit.

In *Problems 8 and 9*, a series connected C–R circuit is suddenly connected to a d.c. source of V volts. Which of the statements are false?

8 (a) The initial current flowing is given by V/R.
(b) The time constant of the circuit is given by CR.
(c) The current grows exponentially.
(d) The final value of the current is zero.

9 (a) The capacitor voltage is equal to the voltage drop across the resistor.
(b) The voltage drop across the resistor decays exponentially.
(c) The initial capacitor voltage is zero.
(d) The initial voltage drop across the resistor is IR, where I is the steady-state current.

10 A capacitor which is charged to V volts is discharged through a resistor of R ohms. Which of the following statements is false?
(a) The initial current flowing is V/R amperes.
(b) The voltage drop across the resistor is equal to the capacitor voltage.
(c) The time constant of the circuit is CR seconds.
(d) The current grows exponentially to a final value of V/R amperes.

An inductor of inductance 0.1 H and negligible resistance is connected in series with a 50 Ω resistor to a 20 V, d.c. supply. In *Problems 11 to 15*, use this data to determine the value required, selecting your answer from those given below.

(a) 5 ms; (b) 12.6 V; (c) 0.4 A; (d) 500 ms; (e) 7.4 V;
(f) 2.5 A; (g) 2 ms; (h) 0 V; (i) 0 A; (j) 20 V.

11 The value of the time constant of the circuit.

12 The approximate value of the voltage across the resistor after a time equal to the time constant.
13 The final value of the current flowing in the circuit.
14 The initial value of the voltage across the inductor.
15 The final value of the steady-state voltage across the inductor.

CONVENTIONAL PROBLEMS

Transients in series connected C–R circuits

1 An uncharged capacitor of 0.2 μF is connected to a 100 V, d.c. supply through a resistor of 100 kΩ. Determine, either graphically or by calculation the capacitor voltage 10 ms after the voltage has been applied. [39.35 V]

2 A circuit consists of an uncharged capacitor connected in series with a 50 kΩ resistor and has a time constant of 15 ms. Determine either graphically or by calculation (a) the capacitance of the capacitor and (b) the voltage drop across the resistor 5 ms after connecting the circuit to a 20 V, d.c. supply. [(a) 0.3 μF; (b) 14.33 V]

3 A 10 μF capacitor is charged to 120 V and then discharged through a 1.5 MΩ resistor. Determine either graphically or by calculation the capacitor voltage 2 s after discharging has commenced. Also find how long it takes for the voltage to fall to 25 V. [105.0 V; 23.53 s]

4 A capacitor is connected in series with a voltmeter of resistance 750 kΩ and a battery. When the voltmeter reading is steady the battery is replaced with a shorting link. If it takes 17 s for the voltmeter reading to fall to two-thirds of its original value, determine the capacitance of the capacitor. [55.9 μF]

5 When a 3 μF charged capacitor is connected to a resistor, the voltage falls by 70% in 3.9 s. Determine the value of the resistor. [1.08 MΩ]

6 A circuit consists of a capacitor and resistor connected in series. When connected to a battery, sketch curves showing the variation of capacitor voltage, voltage drop across the resistor and current flow over the transient period. [See *Fig 2*]

7 An uncharged capacitor is connected in series with a resistor across a d.c. supply. On the same axes sketch curves to show how (a) the current and (b) the capacitor voltage vary with time. [See *Fig 2*]

8 Explain the meaning of the term 'time constant' of a circuit comprising a capacitor and resistor connected in series with a constant voltage d.c. supply. [See para. 3]

9 A 50 μF, uncharged capacitor is connected in series with a 1 kΩ resistor and the circuit is switched to a 100 V, d.c. supply. Determine:
(a) the initial current flowing in the circuit,
(b) the time constant,
(c) the value of current when t is 50 ms and
(d) the voltage across the resistor 60 ms after closing the switch.
[(a) 0.1 A; (b) 50 ms; (c) 36.8 mA; (d) 30.1 V]

10 An uncharged, 5 μF capacitor is connected in series with a 30 kΩ resistor across a 110 V, d.c. supply. Determine the time constant of the circuit and the initial charging current. Use a graphical method to draw the current/time characteristic of the circuit and hence determine the current flowing 120 ms after connecting to the supply.
[150 ms, $3\frac{2}{3}$ mA; 1.65 mA]

11 An uncharged 80 μF capacitor is connected in series with a 1 kΩ resistor and is switched across a 110 V supply. Determine the time constant of the circuit and the initial value of current flowing. Derive graphically the current/time characteristic for the transient condition and hence determine the value of current flowing after (a) 40 ms and (b) 80 ms. [80 ms, 0.11 A; (a) 66.7 mA; (b) 40.5 mA.]

12 A resistor of 0.5 MΩ is connected in series with a 20 μF capacitor and the capacitor is charged to 200 V. The battery is replaced instantaneously by a conducting link. Draw a graph showing the variation of capacitor voltage with time over a period of at least 6 time constants. Determine from the graph the approximate time for the capacitor voltage to fall to 75 V. [9.8 s]

Transients in series connected L-R circuits

13 A coil has an inductance of 1.2 H and a resistance of 40 Ω and is connected to a 200 V d.c. supply. Draw the current/time characteristic and hence determine the approximate value of the current flowing 60 ms after connecting the coil to the supply. [4.3 A]

14 A 25 V d.c. supply is connected to a coil of inductance 1 H and resistance 5 Ω. Use a graphical method to draw the exponential growth curve of current and hence determine the approximate value of the current flowing 100 ms after being connected to the supply. [2 A]

15 An inductor has a resistance of 20 Ω and an inductance of 4 H. It is connected to a 50 V d.c. supply. By drawing the appropriate characteristic find (a) the approximate value of current flowing after 0.1 s and (b) the time for the current to grow to 1.5 A. [(a) 1 A; (b) 0.18 s]

16 A direct voltage is suddenly applied to a coil of resistance R ohms and inductance L henrys. Explain briefly why current does not rise immediately to its steady-state value. [See para. 9]

17 Explain briefly what you understand by the expression 'the time constant of a series connected L–R circuit'. Draw a graph to explain the significance of the term 'constant' and state the value of time constant in terms of L and R. [See para. 10 and worked *Problem 1*]

18 The field winding of a 200 V d.c. machine has a resistance of 20 Ω and an inductance of 500 mH. Calculate:
(a) the time constant of the field winding,
(b) the value of current flow one time constant after being connected to the supply, and
(c) the current flowing 50 ms after the supply has been switched on.
[(a) 25 ms, (b) 6.32 A, (c) 8.65 A]

The effects of circuit time constant on a rectangular wave

19 With the aid of a circuit diagram, explain briefly the effects on the waveform of the capacitor voltage of altering the value of resistance in a series connected C–R circuit, when a rectangular wave is applied to the circuit. What do you understand by the term 'integrator circuit'? [See para. 15]

20 With reference to a rectangular wave applied to a series connected C–R circuit, explain briefly the shape of the waveform when R is small and hence what you understand by the term 'differentiator circuit'. [See para. 16]

6 D.C. machines

A. MAIN POINTS CONCERNED WITH D.C. MACHINES

1 When the input to an electrical machine is electrical energy, (seen as applying a voltage to the electrical terminals of the machine), and the output is mechanical energy, (seen as a rotating shaft), the machine is called an electric **motor**. Thus an electric motor converts electrical energy into mechanical energy.

2 When the input to an electrical machine is mechanical energy, (seen as, say, a diesel motor, coupled to the machine by a shaft), and the output is electrical energy, (seen as a voltage appearing at the electrical terminals of the machine), the machine is called a **generator**. Thus, a generator converts mechanical energy to electrical energy.

3 The efficiency of an electrical machine is the ratio of the output power to the input power and is usually expressed as a percentage. The Greek letter, 'eta', 'η' is used to signify efficiency and since the units are $\frac{\text{power}}{\text{power}}$, then efficiency has no units. Thus

$$\textbf{efficiency, } \boldsymbol{\eta} = \frac{\textbf{output power}}{\textbf{input power}} \times 100\%$$

4 **The action of a commutator** In an electric motor, conductors rotate in a uniform magnetic field. A single-loop conductor mounted between permanent magnets is shown in *Fig 1*. A voltage is applied at points A and B in *Fig 1(a)*.

A force, F, acts on the loop due to the interaction of the magnetic field of the permanent magnets and the magnetic field created by the current flowing in the loop. This force is proportional to the flux density, B, the current flowing, I, and the effective length of the conductor, l, i.e. $F = BIl$. The force is made up of two parts, one acting vertically downwards due to the current flowing from C to D and the other acting vertically upwards due to the current flowing from E to F (from Fleming's left hand rule). If the loop is free to rotate, then when it has rotated through 180°, the conductors are as shown in *Fig 1(b)*. For rotation to continue in the same direction, it is necessary for the current flow to be as shown in *Fig 1(b)*, i.e. from D to C and from F to E. This apparent reversal in the direction of current flow is achieved by a process called **commutation**. With reference to *Fig 2(a)*, when a direct voltage is applied at A and B, then as the single-loop conductor rotates, current flow will always be away from the commutator for the part of the conductor adjacent to the N-pole and towards the commutator for the part of the conductor adjacent to the S-pole. Thus the

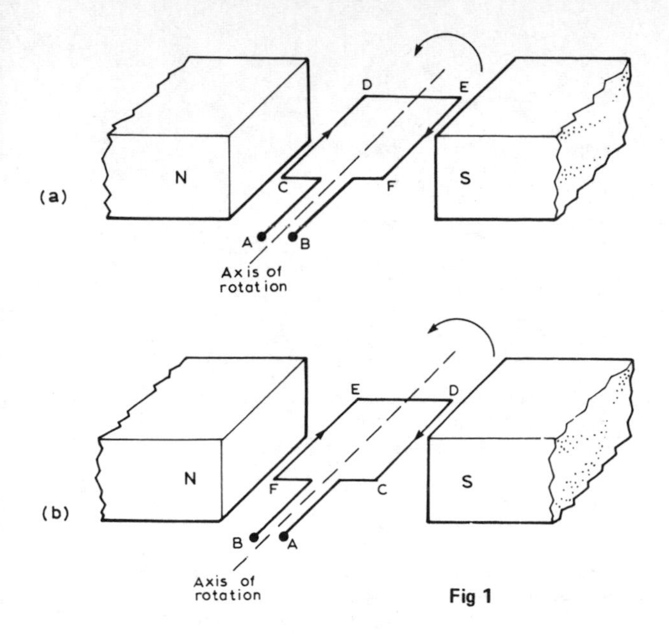

Fig 1

forces act to give continuous rotation in an anti-clockwise direction. The arrangement shown in *Fig 2* is called a 'two-segment' commutator and the voltage is applied to the rotating segments by stationary **brushes**, (usually carbon blocks), which slide on the commutator material, (usually copper), when rotation takes place.

In practice, there are many conductors on the rotating part of a d.c. machine and these are attached to many commutator segments. A schematic diagram of a multi-segment commutator is shown in *Fig 2(b)*.

5 **d.c. machine construction.** The basic parts of any d.c. machine are shown in *Fig 3*, and comprise:

(a) a stationary part called the **stator** having,

(i) a steel ring called the **yoke**, to which are attached

(ii) the magnetic **poles**, around which are the

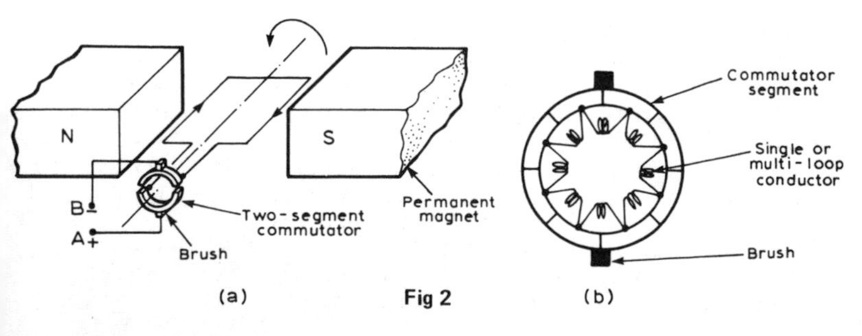

Fig 2

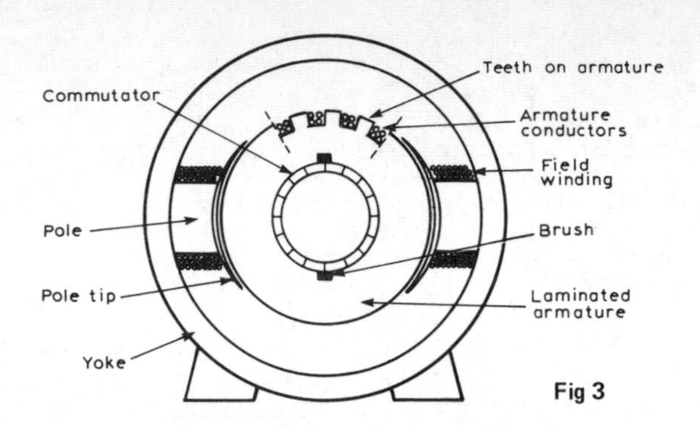

Fig 3

(iii) **field windings**, i.e. many turns of a conductor wound round the pole core. Current passing through this conductor creates an electromagnet, (rather than the permanent magnets shown in *Figs 1 and 2*).

(b) a rotating part called the **armature** mounted in bearings housed in the stator and having,

(iv) a laminated cylinder of iron or steel called the **core**, on which teeth are cut to house the

(v) **armature winding**, i.e. a single or multi-loop conductor system and

(vi) the **commutator**, (see para. 4).

6 The average e.m.f. induced in a single conductor on the armature of a d.c. machine is given by

$$\frac{\text{flux cut/rev}}{\text{time of 1 rev}} = \frac{2p\Phi}{1/n}$$

where p is the number of **pairs** of poles, Φ is the flux in Wb entering or leaving a pole and n is the speed of rotation in rev/s. Thus the average e.m.f. per conductor is $2p\Phi n$ volts. If there are Z conductors connected in series, the average e.m.f. generated is $2p\Phi nZ$ volts. For a given machine, the number of pairs of poles p and the number of conductors connected in series Z are constant, hence the generated e.m.f. is proportional to Φn. But $2\pi n$ is the angular velocity, ω, in rad/s, hence the generated e.m.f. E is proportional to Φ and to ω,

i.e. generated e.m.f., $E \propto \Phi\omega$ (1)

7 The power on the shaft of a d.c. machine is the product of the torque and the angular velocity, i.e.

shaft power $= T\omega$ watts

where T is the torque in N m and ω is the angular velocity in rad/s. The power developed by the armature current is EI_a watts, where E is the generated e.m.f. in volts and I_a is the armature current in amperes. If losses are neglected then $T\omega = EI_a$. But from para. 6, $E \propto \Phi\omega$

Hence $T\omega \propto \Phi\omega I_a$, i.e. $T \propto \Phi I_a$ (2)

8 The principal **losses of machines** are:

(i) **Copper loss,** due to I^2R heat losses in the armature and field windings.

(ii) **Iron (or core) loss,** due to hysteresis and eddy-current losses in the armature. This loss can be reduced by constructing the armature of silicon steel laminations having a high resistivity and low hysteresis loss. At constant speed, the iron loss is assumed constant.

(iii) **Friction and windage losses,** due to bearing and brush contact friction and losses due to air resistance against moving parts (called windage). At constant speed, these losses are assumed to be constant.

(iv) **Brush contact loss** between the brushes and commutator. This loss is approximately proportional to the load current.

The total losses of a machine can be quite significant and operating efficiencies of between 80% and 90% are common.

9 When the field winding of a d.c. machine is connected in parallel with the armature, as shown in *Fig 4(a)*, the machine is said to be **shunt** wound. If the field winding is connected in series with the armature, as shown in *Fig 4(b)*, then the machine is said to be **series** wound.

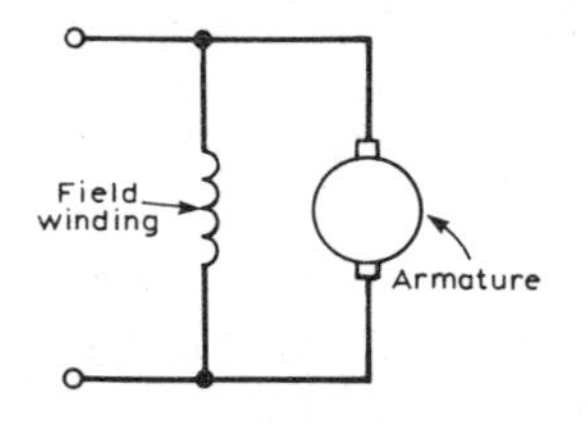

(a) Shunt-wound machine **Fig 4** (b) Series-wound machine

10 Depending on whether the electrical machine is series wound or shunt wound, it behaves differently when a load is applied. The behaviour of a d.c. machine under various conditions is shown by means of graphs, called characteristic curves or just **characteristics**. The characteristics shown in the paras below and in the worked problems are theoretical, since they neglect the effects of such things as armature reaction and demagnetising ampere-turns, which are beyond the scope of this text.

11 **Shunt-wound motor characteristics.** The two principal characteristics are the torque/armature current and speed/armature current relationships. From these, the torque/speed relationship can be derived.

(i) The theoretical torque/armature current characteristic can be derived from the expression $T \propto \Phi I_a$, (see para. 7). For a shunt-wound motor, the field winding is connected in parallel with the armature circuit and thus the applied voltage gives a constant field current, i.e., a shunt-wound motor is a constant flux machine. Since Φ is constant, it follows that $T \propto I_a$, and the characteristic is as shown in *Fig 5(a)*.

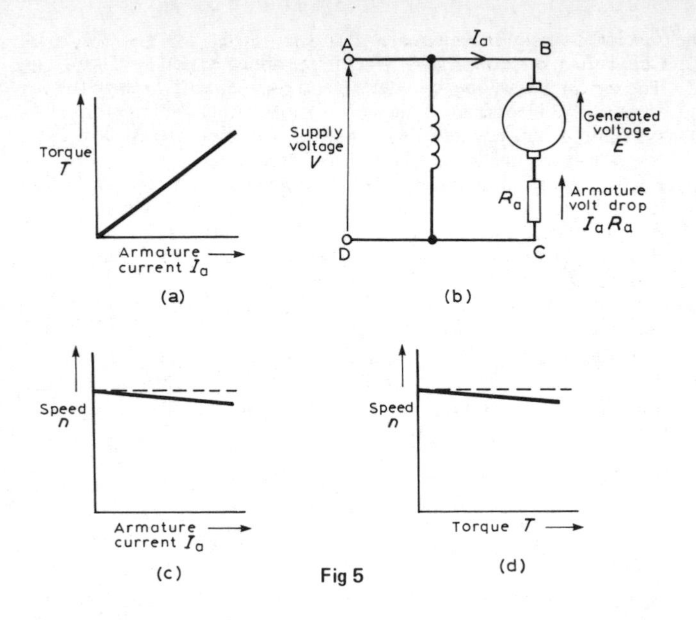

Fig 5

(ii) The armature circuit of a d.c. motor has resistance due to the armature winding and brushes, R_a ohms, and when armature current I_a is flowing through it, there is a voltage drop of I_aR_a volts. In *Fig 5(b)* the armature resistance is shown as a separate resistor in the armature circuit to help understanding. Also, even though the machine is a motor, because conductors are rotating in a magnetic field, a voltage, $E \propto \Phi\omega$, is generated by the armature conductors. By applying Kirchhoff's voltage law to the armature circuit ABCD in *Fig 5(b)*, the voltage equation is $V = E+I_aR_a$, i.e. $E = V-I_aR_a$. But from para. 6, $E \propto \Phi n$, hence $n \propto E/\Phi$, i.e.

$$\text{speed of rotation, } n \propto \frac{E}{\Phi} \propto \frac{V-I_aR_a}{\Phi} \qquad (3)$$

For a shunt motor, V, Φ and R_a are constants, hence as armature current I_a increases, I_aR_a increases and $V-I_aR_a$ decreases, and the speed is proportional

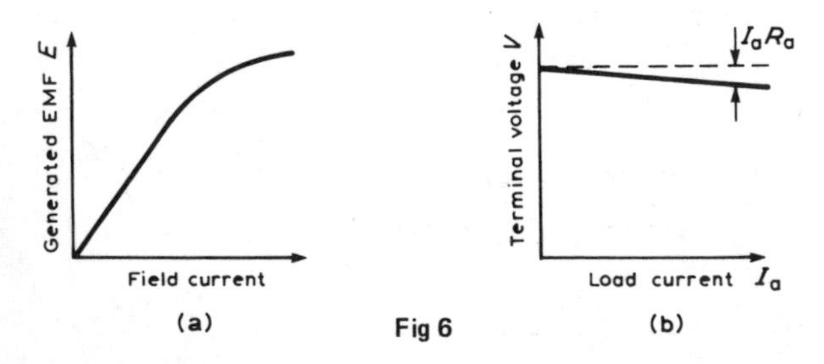

Fig 6

to a quantity which is decreasing and is as shown in *Fig 5(c)*. As the load on the shaft of the motor increases, I_a increases and the speed drops slightly. In practice, the speed falls by about 10% between no-load and full-load on many d.c. shunt-wound motors.

(iii) Since torque is proportional to armature current, (see (i) above), the theoretical speed/torque characteristic is as shown in *Fig 5(d)*.

12 **Series-wound motor characteristics.** The torque/current, speed/current and speed/torque characteristics are discussed in *Problem 7* (page 126) and the characteristics are shown in *Fig 9*.

13 **Shunt-wound generator characteristics.** The two principal generator characteristics are the generated voltage/field current characteristic, called the open-circuit characteristic and the terminal voltage/load current characteristic, called the load characteristic.

(i) **The theoretical open-circuit characteristic.** The generated e.m.f., E, is proportional to $\Phi\omega$, (see para. 6), hence at constant speed, since $\omega = 2\pi n$, $E \propto \Phi$. Also the flux Φ is proportional to field current I_f until magnetic saturation of the iron circuit of the generator occurs. Hence the open circuit characteristic is as shown in *Fig 6(a)*.

(ii) **The theoretical load characteristic.** As the load current on a generator having constant field current and running at constant speed increases, the value of armature current increases, hence the armature volt drop, I_aR_a increases. The generated voltage E is larger than the terminal voltage V and the voltage equation for the armature circuit is $V = E - I_aR_a$. Since E is constant, V decreases with increasing load. The load characteristic is as shown in *Fig 6(b)*. In practice, the fall in voltage is about 10% between no-load and full-load for many d.c. shunt-wound generators.

14 **Series-wound generator characteristic.** The characteristic curve for a series-wound generator is discussed in *Problem 8* (page 126)

15 **The d.c. motor starter.** If a d.c. motor whose armature is stationary is switched directly to its supply voltage, it is likely that the fuses protecting the motor will burn out. This is because the armature resistance is small, frequently being less than one ohm. Thus, additional resistance must be added to the armature circuit at the instant of closing the switch to start the motor.

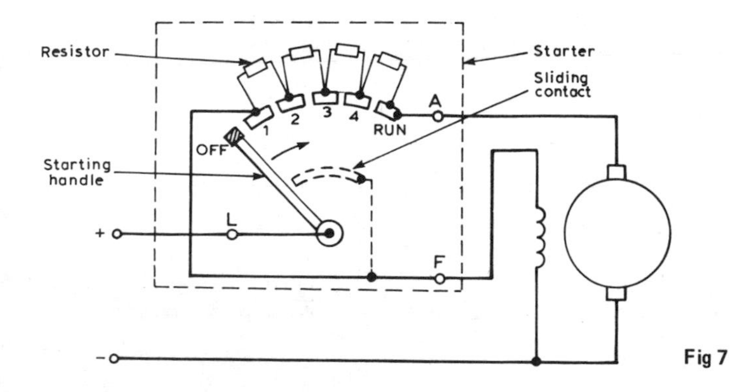

Fig 7

As the speed of the motor increases, the armature conductors are cutting flux and a generated voltage, acting in opposition to the applied voltage, is produced, which limits the flow of armature current. Thus the value of the additional armature resistance can then be reduced.

When at normal running speed, the generated e.m.f. is such that no additional resistance is required in the armature circuit. To achieve this varying resistance in the armature circuit on starting, a d.c. motor starter is used, as shown in *Fig 7*. The starting handle is moved **slowly** in a clockwise direction to start the motor. For a shunt-wound motor, the field winding is connected to stud 1 or to L via a sliding contact on the starting handle, to give maximum field current, hence maximum flux, hence maximum torque on starting, since $T \propto \Phi I_a$. A similar arrangement without the field connection is used for series motors.

16 **Speed control of d.c. motors.**

(i) **Shunt-wound motors.** The speed of a shunt-wound d.c. motor, n, is proportional to $(V - I_aR_a)/\Phi$, (see para. 11). The speed is varied either by varying the value of flux, Φ, or by varying the value of R_a. The former is achieved by using a variable resistor in series with the field winding, as shown in *Fig 8(a)* and such a resistor is called the **shunt field regulator**. As the value of resistance of the shunt field regulator is increased, the value of the field current, I_f, is decreased. This results in a decrease in the value of flux, Φ, and hence an increase in the speed, since $n \propto 1/\Phi$. Thus only speeds **above** that given without a shunt field regulator can be obtained by this method. Speeds **below** those given by $(V-I_aR_a)/\Phi$ are obtained by increasing the resistance in the armature circuit, as shown in *Fig 8(b)*, where

$$n \propto \frac{V-I_a(R_a+R)}{\Phi}.$$

Since resistor R is in series with the armature, it carries the full armature current and results in a large power loss in large motors where a considerable speed reduction is required for long periods.

(ii) **Series-wound motors.** The speed control of series-wound motors is discussed in *Problem 9* (page 127).

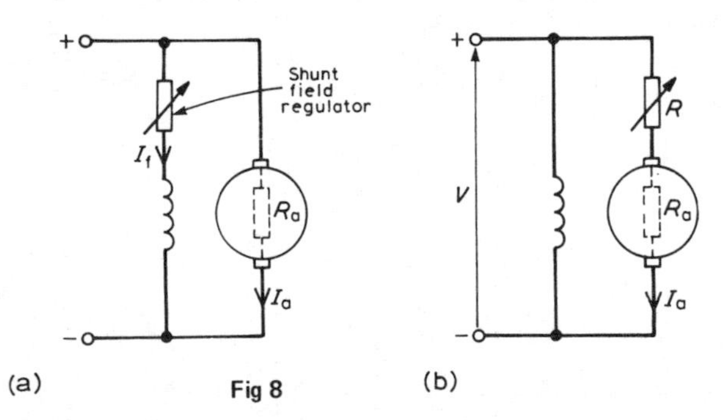

Fig 8

17 **A stepping motor** 'steps' from one fixed position to the next in a sequence of discrete movements – unlike the smooth and continuous rotation of the

machines previously discussed. Instead of having a continuous input signal, digital control with pulses are used.

Stepping motors are used typically

(i) as head positioners in floppy disc drives
(ii) in computer–controlled X–Y plotters
(iii) in teletype printers and
(iv) in robotic applications.

Modern stepping motors operate on the principle that various combinations and phasings of the fields, and the interaction between these fields and the rotor field within the motors causes the rotor to move a defined number of degrees – either forward or backwards. The number of pulses and the frequency of the control signal determines the number of steps and the speed of rotation. A step angle of 1.8° is common, thus giving $\frac{360}{1.8}$, i.e., 200 steps per revolution.

B. WORKED PROBLEMS ON D.C. MACHINES

Problem 1 A 200 V d.c. motor develops a shaft torque of 15 Nm at 1200 rev/min. If the efficiency is 80%, determine the current supplied to the motor.

From para. 3, the efficiency of a motor $= \frac{\text{output power}}{\text{input power}} \times 100\%$.

The output power of a motor is the power available to do work at its shaft and is given by $T\omega$ or $T(2\pi n)$ watts, where T is the torque in Nm and n is the speed of rotation in rev/s. The input power is the electrical power in watts supplied to the motor, i.e. VI watts.

Thus for a motor, efficiency, $\eta = \frac{T(2\pi n)}{VI} \times 100\%$

i.e., $$80 = \left\{ \frac{(15)(2\pi)(\frac{1200}{60})}{(200)(I)} \right\} (100)$$

Thus the current supplied, $$I = \frac{(15)(2\pi)(20)(100)}{(200)(80)} = 11.8 \text{ A}$$

Problem 2 A 100 V d.c. generator supplies a current of 15 A when running at 1500 rev/min. If the torque on the shaft driving the generator is 12 Nm, determine (a) the efficiency of the generator and (b) the power loss in the generator.

(a) From para. 3, the efficiency of a generator $= \frac{\text{output power}}{\text{input power}} \times 100\%$

The output power is the electrical output, i.e. VI watts. The input power to a generator is the mechanical power in the shaft driving the generator, i.e. $T\omega$ or $T(2\pi n)$ watts, where T is the torque in Nm and n is speed of rotation in rev/s. Hence, for a generator

$$\text{efficiency,} \quad \eta = \frac{VI}{T(2\pi n)} \times 100\%$$

$$\text{i.e.} \quad \eta = \frac{(100)(15)(100)}{(12)(2\pi)(\frac{1500}{60})}$$

i.e. **efficiency** = **79.6%**.

(b) The input power = output power + losses

Hence, $T(2\pi n) = VI + \text{losses}$

i.e. losses $= T(2\pi n) - VI$

$$= \left[(12)(2\pi)(\frac{1500}{60})\right] - \left[(100)(15)\right]$$

i.e. **power loss** = 1885–1500 = **385 W**

Problem 3 A d.c. shunt-wound generator running at constant speed generates a voltage of 150 V at a certain value of field current. Determine the change in the generated voltage when the field current is reduced by 20%, assuming the flux is proportional to the field current.

The generated e.m.f. E of a generator is proportional to $\Phi\omega$, (see para. 6), i.e. is proportional to Φn, where Φ is the flux and n is the speed of rotation. It follows that $E = k\Phi n$, where k is a constant. At speed n_1 and flux Φ_1, $E_1 = k\Phi_1 n_1$. At speed n_2 and flux ϕ_2, $E_2 = k\Phi_2 n_2$. Thus, by division:

$$\frac{E_1}{E_2} = \frac{k\Phi_1 n_1}{k\Phi_2 n_2} = \frac{\Phi_1 n_1}{\Phi_2 n_2}$$

The initial conditions are $E_1 = 150$ V, $\Phi = \Phi_1$ and $n = n_1$. When the flux is reduced by 20%, the new value of flux is 80/100 of 0.8 of the initial value, i.e. $\Phi_2 = 0.8\Phi_1$. Since the generator is running at constant speed, $n_2 = n_1$. Thus

$$\frac{E_1}{E_2} = \frac{\Phi_1 n_1}{\Phi_2 n_2} = \frac{\Phi_1 n_1}{0.8\Phi_1 n_1} = \frac{1}{0.8}$$

that is, $E_2 = 150 \times 0.8 = 120$ V.

Thus, a reduction of 20% in the value of the flux **reduces the generated voltage to 120 V** at constant speed.

Problem 4 A 200 V, d.c. shunt-wound motor has an armature resistance of 0.4 Ω and at a certain load has an armature current of 30 A and runs at 1350 rev/min. If the load on the shaft of the motor is increased so that the armature current increases to 45 A, determine the speed of the motor, assuming the flux remains constant.

The relationship $E \propto \Phi n$ applies to both generators and motors. For a motor,

$E = V - I_a R_a$, (see para. 11).

Hence $E_1 = 200 - 30 \times 0.4 = 188$ V,

and $E_2 = 200 - 45 \times 0.4 = 182$ V.

With reference to *Problem 3*, the relationship

$$\frac{E_1}{E_2} = \frac{\Phi_1 n_1}{\Phi_2 n_2}$$

applies to both generators and motors. Since the flux is constant, $\Phi_1 = \Phi_2$.

Hence $\dfrac{188}{182} = \dfrac{\Phi_1 \times \dfrac{1350}{60}}{\Phi_1 \times n_2}$, i.e., $n_2 = \dfrac{22.5 \times 182}{188}$

$= 21.78$ rev/s.

Thus the speed of the motor when the armature current is 45 A is 21.78 × 60 rev/min, i.e. **1307 rev/min.**

Problem 5 The shaft torque of a diesel motor driving a 100 V d.c., shunt-wound generator is 25 Nm. The armature current of the generator is 16 A at this value of torque. If the shunt field regulator is adjusted so that the flux is reduced by 15%, the torque increases to 35 Nm. Determine the armature current at this new value of torque.

The shaft torque T of a generator is proportional to ΦI_a, where Φ is the flux and I_a is the armature current. Thus, $T = k\Phi I_a$ where k is a constant. The torque at flux Φ_1 and armature current I_{al} is $T_1 = k\Phi_1 I_{al}$. Similarly, $T_2 = k\Phi_2 I_{a2}$.

By division $\dfrac{T_1}{T_2} = \dfrac{k\Phi_1 I_{al}}{k\Phi_2 I_{a2}} = \dfrac{\Phi_1 I_{al}}{\Phi_2 I_{a2}}$

Hence $\dfrac{25}{35} = \dfrac{\Phi_1 \times 16}{0.85\Phi_1 \times I_{a2}}$

i.e. $I_{a2} = \dfrac{16 \times 35}{0.85 \times 25} = 26.35$ A

That is, **the armature current at the new value of torque is 26.35 A.**

Problem 6 A 220 V, d.c. shunt-wound motor runs at 800 rev/min and the armature current is 30 A. The armature circuit resistance is 0.4 Ω. Determine (a) the maximum value of armature current if the flux is suddenly reduced by 10% and (b) the steady state value of the armature current at the new value of flux, assuming the shaft torque of the motor remains constant.

(a) For a d.c. shunt-wound motor, $E = V - I_a R_a$. Hence initial generated e.m.f., $E_1 = 220 - 30 \times 0.4 = 208$ V. The generated e.m.f. is also such that $E \propto \Phi n$, so at the instant the flux is reduced, the speed has not had time to change, and

$E = 208 \times \dfrac{90}{100} = 187.2$ V.

Hence, the voltage drop due to the armature resistance is 220–187.2, i.e., 32.8 V. The **instantaneous value of the current** is 32.8/0.4, i.e. **82 A.** This increase in current is about three times the initial value and causes an increase in torque, ($T \propto \Phi I_a$). The motor accelerates because of the larger torque value until steady state conditions are reached.

(b) $T \propto \Phi I_a$ and since the torque is constant,
$\Phi_1 I_{al} = \Phi_2 I_{a2}$. The flux Φ is reduced by 10%, hence
$\Phi_2 = 0.9\Phi_1$.

Thus, $\Phi_1 \times 30 = 0.9\Phi_1 \times I_{a2}$

i.e. the steady state value of armature current, $I_{a2} = \frac{30}{0.9} = 33\frac{1}{3}$ A.

Problem 7 Sketch the torque/current, speed/current and speed/torque characteristics for a d.c., series-wound motor and with reference to e.m.f. and torque relationships for a d.c. machine, explain their shape.

In a series motor, the armature current flows in the field winding and is equal to the supply current I, (see *Fig 4(b)*).

(i) **The torque/current characteristic.** It is shown in para. 7 that torque $T \propto \Phi I_a$. Since the armature and field currents are the same current, I, in a series machine, then $T \propto \Phi I$ over a limited range, before magnetic saturation of the magnetic circuit of the motor is reached, (i.e., the linear portion of the B–H curve for the yoke, poles, air gap, brushes and armature in series). Thus $\Phi \propto I$ and $T \propto I^2$. After magnetic saturation, Φ almost becomes a constant and

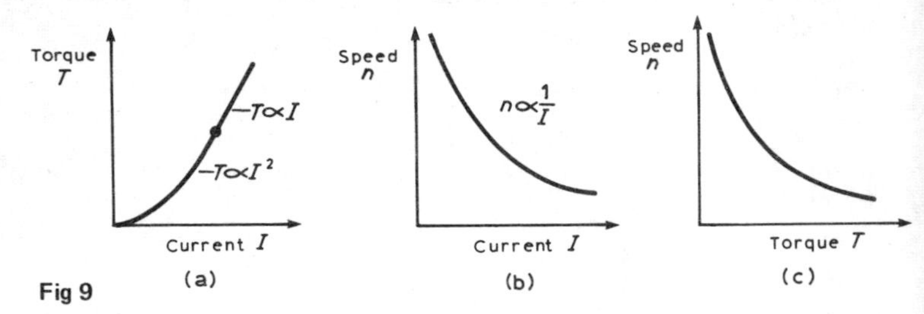

Fig 9

$T \propto I$. Thus the theoretical torque/current characteristic is as shown in *Fig 9(a)*.

(ii) **The speed/current characteristic.** It is shown in para. 11(ii) that $n \propto (V - I_a R_a)/\Phi$. In a series motor, $I_a = I$ and below the magnetic saturation level, $\Phi \propto I$. Thus $n \propto (V - IR)/I$ where R is the combined resistance of the series field and armature circuit. Since IR is small compared with V, then an approximate relationship for the speed is $n \propto 1/I$. Hence the theoretical speed/current characteristic is as shown in *Fig 9(b)*. The high speed at small values of current indicate that this type of motor must not be run on very light loads and invariably, such motors are permanently coupled to their loads.

(iii) The theoretical speed/torque characteristic may be derived from (i) and (ii) above by obtaining the torque and speed for various values of current and plotting the co-ordinates on the speed/torque characteristic. A typical speed/torque characteristic is shown in *Fig 9(c)*.

Problem 8 Sketch the theoretical load characteristic for a series-wound generator and explain its shape with reference to the generated e.m.f. relationship for a d.c. machine. Explain why it is not possible to obtain an open-circuit characteristic for a series-wound generator.

The load characteristic is the terminal voltage/current characteristic. The generated e.m.f., E, is proportional to $\Phi\omega$ and at constant speed ω ($= 2\pi n$) is a constant. Thus E is proportionai to Φ. For values of current below magnetic saturation of the yoke, poles, air gaps and armature core, the flux Φ is proportional to the current, hence $E \propto I$. For values of current above those required for magnetic saturation, the generated e.m.f. is approximately constant. The values of field resistance and armature resistance in a series wound machine are small, hence the terminal voltage V is very nearly equal to E. Thus the theoretical load characteristic is similar in shape to the characteristic shown in *Fig 6(a)*.

In a series-wound generator, the field winding is in series with the armature and it is not possible to have a value of field current when the terminals are open circuited, thus it is not possible to obtain an open-circuit characteristic.

Problem 9 Explain how the speed of a d.c. series-wound motor can be controlled by using (a) field resistance and (b) armature resistance techniques.

(a) The speed of a d.c. series-wound motor is given by:

$$n = k\left(\frac{V - IR}{\Phi}\right)$$

where k is a constant, V is the terminal voltage, R is the combined resistance of the armature and series field and Φ is the flux.

Thus, a reduction in flux results in an increase in speed. This is achieved by putting a variable resistance in parallel with the field winding and reducing the field current, and hence flux, for a given value of supply current. A circuit diagram of this arrangement is shown in *Fig 10(a)*. A variable resistor connected in parallel with the series-wound field to control speed is called a **diverter**. Speeds above those given with no diverter are obtained by this method.

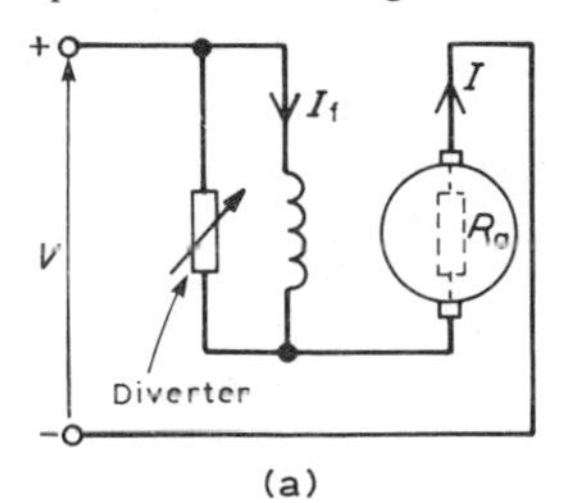

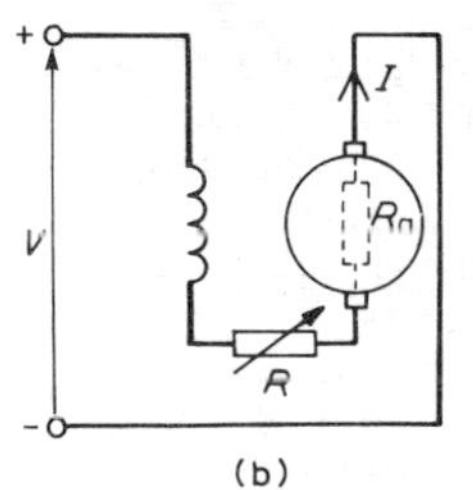

Fig 10

(b) Speeds below normal are obtained by connecting a variable resistor in series with the field winding and armature circuit, as shown in *Fig 10(b)*. This effectively increases the value of R in the equation

$$n = k\left(\frac{V - IR}{\Phi}\right)$$

and thus reduces the speed. Since the additional resistor carries the full supply current, a large power loss is associated with large motors in which a considerable speed reduction is required for long periods.

C. FURTHER PROBLEMS ON D.C. MACHINES

SHORT ANSWER PROBLEMS

1 An electric .... converts electrical energy to mechanical energy.
2 An electric motor converts .... energy to .... energy.

3 The efficiency of an electrical machine is given by the ratio $\frac{\ldots\ldots\ldots\ldots\ldots}{\ldots\ldots\ldots\ldots\ldots}$ %.

In *Problems 4 to 7*, an electrical machine runs at n rev/s, has a shaft torque of T and takes a current of I from a supply of voltage V.

4 The power input to a generator is . . . . watts.
5 The power input to a motor is . . . . watts.
6 The power output from a generator is . . . . watts.
7 The power output from a motor is . . . . .watts.
8 The generated e.m.f. of a d.c. machine is proportional to . . . . volts.
9 The torque produced by a d.c. motor is proportional to . . . . Nm.
10 In a series-wound d.c. machine, the field winding is in . . . . with the armature circuit.
11 A d.c. motor has its field winding in parallel with the armature circuit. It is called a . . . . wound motor.
12 In a d.c. generator, the relationship between the generated voltage, terminal voltage, current and armature resistance is given by $E =$ . . . .
13 The equation relating the generated e.m.f., terminal voltage, armature current and armature resistance for a d.c. motor is . . . . . . = ...
14 A starter is necessary for a d.c. motor because the generated e.m.f. is . . . . at low speeds.
15 The speed of a d.c. shunt-wound motor will . . . . if the value of resistance of the shunt field regulator is increased.
16 The speed of a d.c. motor will . . . . if the value of resistance in the armature circuit is increased.
17 The value of the speed of a d.c. shunt-wound motor . . . . as the value of the armature current increases.
18 At a large value of torque, the speed of a d.c. series-wound motor is . . . .
19 At a large value of field current, the generated e.m.f. of a d.c. shunt-wound generator is approximately . . . .
20 In a series-wound generator, the terminal voltage increases as the load current . . . .

MULTI-CHOICE PROBLEMS (answers on page 191)

1 Which of the following statements is false?
(a) A d.c. motor converts electrical energy to mechanical energy.
(b) The efficiency of a d.c. motor is the ratio $\frac{\text{input power}}{\text{output power}} \times 100\%$
(c) A d.c. generator converts mechanical energy to electrical energy.
(d) The efficiency of a d.c. generator is the ratio $\frac{\text{output power}}{\text{input power}} \times 100\%$

A shunt-wound d.c. machine is running at n rev/s and has a shaft torque of T Nm. The supply current is I A when connected to d.c. bus bars of voltage V volts. The armature resistance of the machine is R_a ohms, the armature current is I_a A and the generated voltage is E volts. Use this data to find the equations of the quantities stated in *Problems 2 to 9*, selecting the correct answer from the list given below.

(a) $V-I_aR_a$; (b) $E+I_aR_a$; (c) VI;
(d) $E-I_aR_a$; (e) $T(2\pi n)$; (f) $V+I_aR_a$.

2 The input power when running as a generator.
3 The output power when running as a motor.
4 The input power when running as a motor.
5 The output power when running as a generator.
6 The generated voltage when running as a motor.
7 The terminal voltage when running as a generator.
8 The generated voltage when running as a generator.
9 The terminal voltage when running as a motor.
10 Which of the following statements is false?
   (a) A commutator is necessary as part of a d.c. motor to keep the armature rotating in the same direction.
   (b) A commutator is necessary as part of a d.c. generator to produce a uni-directional voltage at the terminals of the generator.
   (c) The field winding of a d.c. machine is housed in slots on the armature.
   (d) The brushes of a d.c. machine are usually made of carbon and do not rotate with the armature.
11 If the speed of a d.c. machine is doubled and the flux remains constant, the generated e.m.f.
   (a) remains the same; (b) is doubled; (c) is halved.
12 If the flux per pole of a shunt-wound d.c. generator is increased, and all other variables are kept the same, the speed
   (a) decreases; (b) stays the same; (c) increases.
13 If the flux per pole of a shunt-wound d.c. generator is halved, the generated e.m.f. at constant speed
   (a) is doubled; (b) is halved; (c) remains the same.
14 In a series-wound generator running at constant speed, as the load current increases, the terminal voltage
   (a) increases; (b) decreases; (c) stays the same.
15 Which of the following statements is false for a series-wound d.c. motor?
   (a) The speed decreases with increase of resistance in the armature circuit.
   (b) The speed increases as the flux decreases.
   (c) The speed can be controlled by a diverter.
   (d) The speed can be controlled by a shunt field regulator.

CONVENTIONAL PROBLEMS

1 A 250 V, series-wound motor is running at 500 rev/min and its shaft torque is 130 N m. If its efficiency at this load is 88%, find the current taken from the supply. [30.94 A]
2 In a test on a d.c. motor, the following data was obtained.
   Supply voltage: 500 V. Current taken from the supply: 42.4 A.
   Speed: 850 rev/min. Shaft torque: 187 N m.
   Determine the efficiency of the motor correct to the nearest 0.5% [78.5%]
3 (a) State the principal losses in d.c. machines.
   (b) The shaft torque required to drive a d.c. generator is 18.7 N m when it is running at 1250 rev/min. If its efficiency is 87% under these conditions and the armature current is 17.3 A, determine the voltage at the terminals of the generator. [123.1 V]
4 A 220 V, d.c. generator supplies a load of 37.5 A and runs at 1550 rev/min.

Determine the shaft torque of the diesel motor driving the generator, if the generator efficiency is 78%. [65.2 N m]

5 Describe the need for a commutator on the armature of a d.c. generator, using diagrams to illustrate your answer.

6 Explain the function of a commutator on the armature of a d.c. motor.

7 Draw a labelled diagram showing a cross-section of a two-pole d.c. machine. Describe the functions performed by the field windings, the armature, the commutator and the brushes.

8 Determine the generated e.m.f. of a d.c. machine if the armature resistance is 0.1 Ω and it (a) is running as a motor connected to a 230 V supply, the armature current being 60 A, and (b) is running as a generator with a terminal voltage of 230 V, the armature current being 80 A. [(a) 224 V; (b) 238 V]

9 A d.c. motor has a speed of 900 rev/min when connected to a 460 V supply. Find the approximate value of the speed of the motor when connected to a 200 V supply, assuming the flux decreases by 30% and neglecting the armature volt drop. [559 rev/min]

10 A d.c. generator has a generated e.m.f. of 210 V when running at 700 rev/min and the flux per pole is 120 mWb. Determine the generated e.m.f.
(a) at 1050 rev/min, assuming the flux remains constant,
(b) if the flux is reduced by one-sixth at constant speed, and
(c) at a speed of 1155 rev/min and a flux of 132 mWb.
[(a) 315 V; (b) 175 V; (c) 381.2 V]

11 A 250 V d.c. shunt-wound generator has an armature resistance of 0.1 Ω. Determine the generated e.m.f. when the generator is supplying 50 kW, neglecting the field current of the generator. [270 V]

12 A series-wound motor is connected to a d.c. supply and develops full-load torque when the current is 30 A and speed is 1000 rev/min. If the flux per pole is proportional to the current flowing, find the current and speed at half full-load torque, when connected to the same supply. [21.2 A, 1415 rev/min]

13 Sketch the theoretical speed/torque characteristic for (a) a series-wound, d.c. motor and (b) a shunt-wound, d.c. motor. Use the e.m.f. and torque relationships to explain their shapes.

14 As the current supplied by a d.c. shunt-wound generator increases, the terminal voltage falls. Explain why the voltage falls and sketch the theoretical load characteristic for this generator.

15 Explain the effect of the generated e.m.f. of a d.c. motor and why a d.c. motor starter is necessary.

16 Explain why a d.c. shunt-wound motor needs a starter when connected to a constant-voltage supply and make a sketch of such a starter.

17 One type of d.c. motor uses resistance in series with the field winding to obtain speed variations and another type uses resistance in parallel with the field winding for the same purpose. Explain why these two distinct methods are used and why the field current plays a significant part in controlling the speed of a d.c. motor.

18 Explain the principle of a stepping motor.

7 Introduction to three-phase induction motors

A. MAIN POINTS CONCERNED WITH AN INTRODUCTION TO THREE-PHASE INDUCTION MOTORS

1 In d.c. motors, introduced in chapter 6, conductors on a rotating armature pass through a stationary magnetic field. In a three-phase induction motor, the magnetic field rotates and this has the advantage that no external electrical connections to the rotor need be made.

The result is a motor which: (i) is cheap and robust, (ii) is explosion proof, due to the absence of a commutator or slip-rings and brushes with their

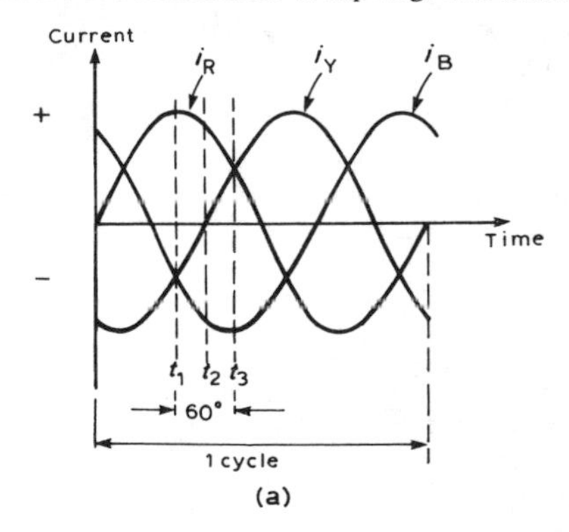

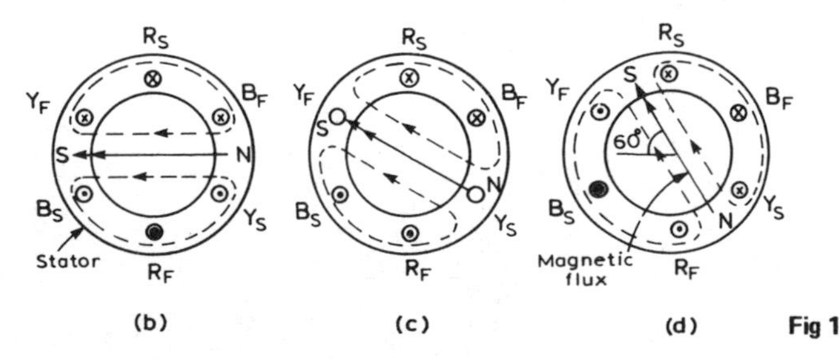

Fig 1

associated sparking, (iii) requires little or no skilled maintenance, and (iv) has self starting properties when switched to a supply with no additional expenditure on auxiliary equipment. The principal disadvantage of a three-phase induction motor is that its speed cannot be readily adjusted.

2 **Production of a rotating magnetic field.** When a three-phase supply is connected to symmetrical three-phase stator windings, the currents flowing in the windings produce a magnetic field. This magnetic field is constant in magnitude and rotates at constant speed as shown below and in *Problem 1*, and is called the **synchronous speed.**

With reference to *Fig 1*, the windings are represented by three single-loop conductors, one for each phase, marked $R_S R_F$, $Y_S Y_F$ and $B_S B_F$, the S and F signifying start and finish. In practice, each phase winding comprises many turns and is distributed around the stator; the single-loop approach is for clarity only.

When the stator windings are connected to a three-phase supply, the current flowing in each winding varies with time and is as shown in *Fig 1(a)*. If the value of current in a winding is positive, the assumption is made that it flows from start to finish of the winding, i.e., if it is the red-phase, current flows from R_S to R_F, i.e. away from the viewer in R_S and towards the viewer in R_F. When the value of current is negative, the assumption is made that it flows from finish to start, i.e. towards the viewer in an 'S' winding and away from the viewer in an 'F' winding.

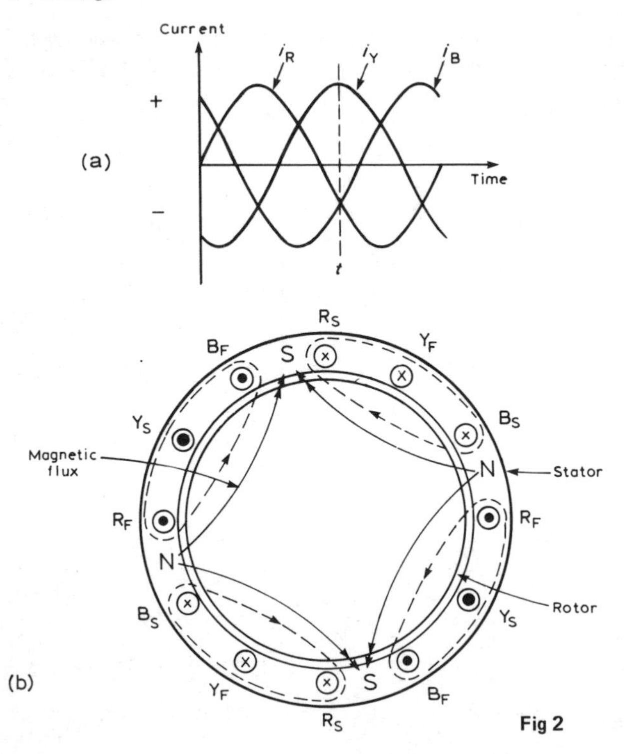

Fig 2

At time, say t_1, shown in *Fig 1(a)*, the current flowing in the red phase is a maximum positive value. At the same time, t_1, the currents flowing in the yellow and blue phases are both 0.5 times the maximum value and are negative. The current distribution in the stator windings is therefore as shown in *Fig 1(b)*, in which current flows away from the viewer, (shown as X) in R_S since it is positive,. but towards the viewer (shown as •) in Y_S and B_S, since these are negative. The resulting magnetic field is as shown, due to the 'solenoid' action and application of the corkscrew rule.

A short time later at time t_2, the current flowing in the red phase has fallen to about 0.87 times its maximum value and is positive, the current in the yellow phase is zero and the current in the blue phase is about 0.87 times its maximum value and is negative. Hence the currents and resultant magnetic field are as shown in *Fig 1(c)*. At time t_3, the currents in the red and yellow phases are 0.5 of their maximum values and the current in the blue phase is a maximum negative value. The currents and resultant magnetic field are as shown in *Fig 1(d)*.

Similar diagrams to *Fig 1(b), (c) and (d)* can be produced for all time values and these would show that the magnetic field travels through one revolution for each cycle of the supply voltage applied to the stator windings. By considering the flux values rather than the current values, it can be shown that the rotating magnetic field has a constant value of flux, (see *Problem 1*).

3 The rotating magnetic field produced by three phase windings could have been produced by rotating a permanent magnet's north and south pole at synchronous speed, (shown as N and S at the ends of the flux phasors in *Figs 1(b), (c)* and *(d)*). For this reason, it is called a 2-pole system and an induction motor using three phase windings only is called a 2-pole induction motor.

If six windings displaced from one another by 60° are used, as shown in *Fig 2(b)*, by drawing the current and resultant magnetic field diagrams at various time values, it may be shown that one cycle of the supply current to the stator windings causes the magnetic field to move through half a revolution. The current distribution in the stator windings are shown in *Fig 2(b)*, for the time t shown in *Fig 2(a)*.

It can be seen that for six windings on the stator, the magnetic flux produced is the same as that produced by rotating two permanent magnet north poles and two permanent magnet south poles at synchronous speed. This is called a 4-pole system and an induction motor using six phase windings is called a 4-pole induction motor. By increasing the number of phase windings the number of poles can be increased to any even number.

In general, if f is the frequency of the currents in the stator windings and the stator is wound to be equivalent to p **pairs** of poles, the speed of revolution of the rotating magnetic field, i.e., the synchronous speed, n_s is given by:

$$n_s = \frac{f}{p} \text{ rev/s}$$

4 **The principle of operation of the three-phase induction motor**. The stator of a three-phase induction motor is the stationary part corresponding to the yoke of a d.c. machine. It is wound to give a 2-pole, 4-pole, 6-pole, ... rotating magnetic field, depending on the rotor speed required. The rotor, corresponding to the armature of a d.c. machine, is built up of laminated iron, to reduce eddy currents.

In the type most widely used, known as a **squirrel-cage rotor**, copper or aluminium bars are placed in slots cut in the laminated iron, the ends of the bars

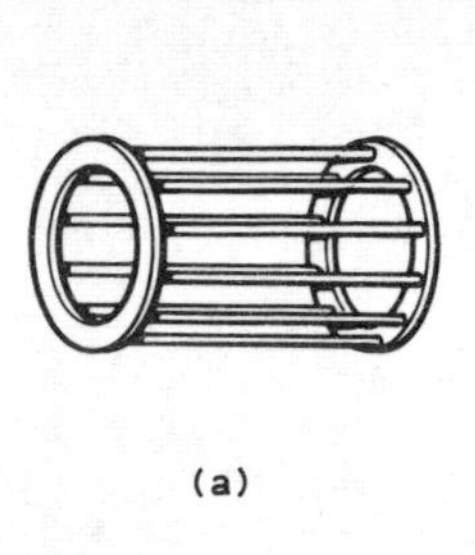

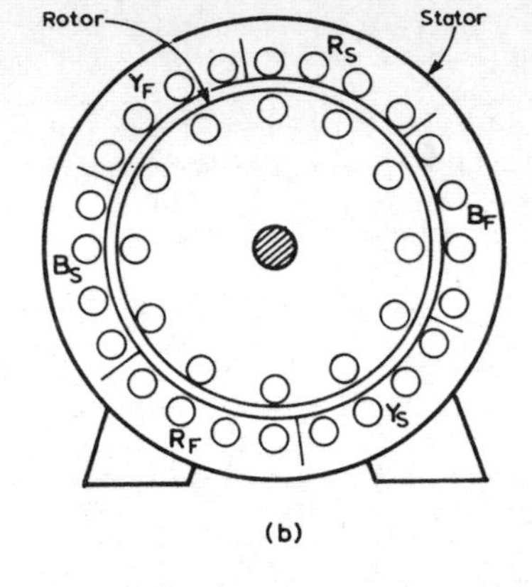

Fig 3

being welded or brazed into a heavy conducting ring, (see *Fig 3(a)*). A cross-sectional view of a three-phase induction motor is shown in *Fig 3(b)*.

When a three-phase supply is connected to the stator windings, a rotating magnetic field is produced. As the magnetic flux cuts a bar on the rotor, an e.m.f. is induced in it and since it is joined, via the end conducting rings, to another bar one pole pitch away, a current flows in the bars. The magnetic field associated with this current flowing in the bars interacts with the rotating magnetic field and a force is produced, tending to turn the rotor in the same direction as the rotating magnetic field, (see *Fig 4*).

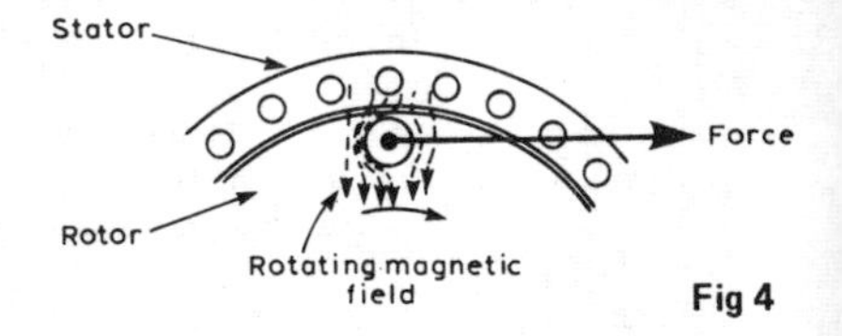

Fig 4

5 **Slip.** The force exerted by the rotor bars causes the rotor to turn in the direction of the rotating magnetic field. As the rotor speed increases, the rate at which the rotating magnetic field cuts the rotor bars is less and the frequency of the induced emf's in the rotor bars is less. If the rotor runs at the same speed as the rotating magnetic field, no emf's are induced in the rotor, hence there is no force on them and no torque on the rotor. Thus the rotor slows down. For this reason the rotor can never run at synchronous speed.

When there is no load on the rotor, the resistive forces due to windage and bearing friction are small and the rotor runs very nearly at synchronous speed. As the rotor is loaded, the speed falls and this causes an increase in the frequency of the induced emf's in the rotor bars and hence the rotor current, force and torque increase. The difference between the rotor speed, n_r and the synchronous speed, n_s, is called the **slip speed**, i.e.

slip speed $= n_s - n_r$ **rev/s**

The ratio $\frac{n_s - n_r}{n_s}$ is called the **fractional slip** or just the **slip**, s, and is usually expressed as a percentage. Thus

slip, $s = \frac{n_s - n_r}{n_s} \times 100\%$

Typical values of slip between no load and full load are about 4 to 5% for small motors and 1½ to 2% for large motors.

B. WORKED PROBLEMS ON AN INTRODUCTION TO THREE-PHASE INDUCTION MOTORS

Problem 1 Show, by means of diagrams, that a rotating magnetic field of constant magnitude can be produced by applying a three-phase supply to three similar coils, displaced from one another by 120°.

The three coils shown in *Fig 5(a)*, are connected in star to a three-phase supply. Let the positive directions of the fluxes produced by currents flowing in the coils, be ϕ_A, ϕ_B and ϕ_C respectively. The directions of ϕ_A, ϕ_B and ϕ_C do not alter, but their magnitudes are proportional to the currents flowing in the coils at any particular time. At time t_1, shown in *Fig 5(b)*, the currents flowing in the coils are:
i_B, a maximum positive value, i.e., the flux is towards point P;
i_A and i_C, half the maximum value and negative, i.e. the flux is away from point P.

These currents give rise to the magnetic fluxes ϕ_A, ϕ_B and ϕ_C, whose magnitudes and directions are as shown in *Fig 5(c)*. The resultant flux is the phasor sum of ϕ_A, ϕ_B and ϕ_C, shown as Φ in *Fig 5(c)*. At time t_2, the currents flowing are:
i_B, 0.866 × maximum positive value; i_C, zero and
i_A, 0.866 × maximum negative value.
The magnetic fluxes and the resultant magnetic flux are as shown in *Fig 5(d)*.
At time t_3,
i_B is 0.5 × maximum value and is positive

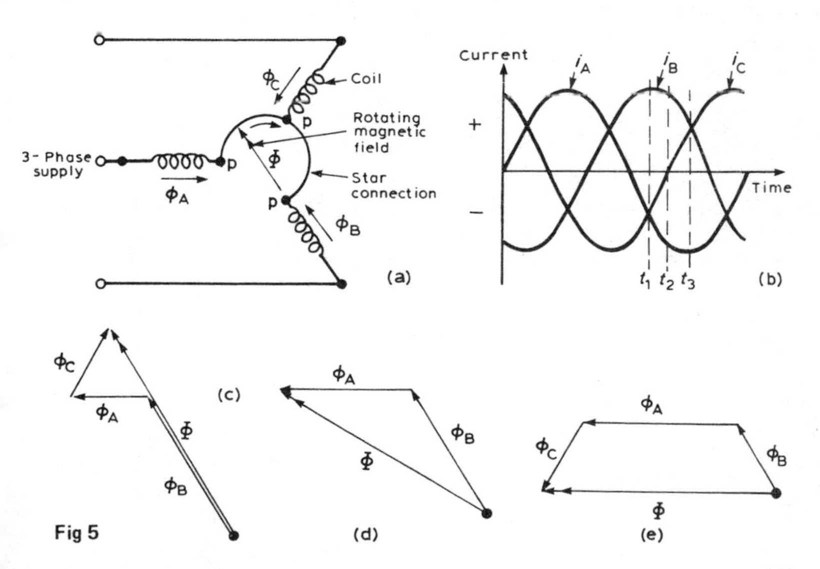

Fig 5

i_A is a maximum negative value, and
i_C is 0.5 × maximum value and is positive.
The magnetic fluxes and the resultant magnetic flux are as shown in *Fig 5(e)*.

Inspection of *Figs 5(c), (d) and (e)* shows that the magnitude of the resultant magnetic flux, Φ, in each case is constant and is 1½ × the maximum value of ϕ_A, ϕ_B or ϕ_C, but that its direction is changing. The process of determining the resultant flux may be repeated for all values of time and shows that the magnitude of the resultant flux is constant for all values of time and also that it rotates at constant speed, making one revolution for each cycle of the supply voltage.

Problem 2 The stator of a 3-phase induction motor is connected to a 50 Hz supply. If the stator is wound to give a 2-pole system, find the synchronous speed of the motor in rev/min.

From para. 3, $n_s = f/p$ rev/s, where n_s is the synchronous speed, f is the frequency in hertz of the supply to the stator and p is the number of **pairs** of poles. Since the stator is connected to a 50 hertz supply, $f = 50$. The motor has a two-pole system, hence p, the number of pairs of poles is one. Thus

synchronous speed, $n_s = \frac{50}{1} = 50$ rev/s $= 50 \times 60$ rev/min $=$ **3000 rev/min.**

Problem 3 The stator of a 3-phase, 4-pole induction motor is connected to a 50 Hz supply. The rotor runs at 1455 rev/min at full load. Determine (a) the synchronous speed and (b) the slip at full load.

(a) The number of pairs of poles, $p = 4/2 = 2$. The supply frequency $f = 50$ Hz.

The **synchronous speed** $n_s = \frac{f}{p} = \frac{50}{2} =$ **25 rev/s**

(b) The rotor speed $n_r = \frac{1455}{60} = 24.25$ rev/s

The slip s $= \frac{n_s - n_r}{n_s} \times 100\%$

$= \frac{25 - 24.25}{25} \times 100\% =$ **3%**

Problem 4 A 3-phase, 60 Hz induction motor has 2 poles. If the slip is 2% at a certain load, determine (a) the synchronous speed, (b) the speed of the rotor and (c) the frequency of the induced e.m.f.'s in the rotor.

(a) $f = 60$ Hz, $p = \frac{2}{2} = 1$.

Hence **synchronous speed,** $n_s = \frac{f}{p} = \frac{60}{1} =$ **60 rev/s**

(b) Since slip, $s = \frac{n_s - n_r}{n_s} \times 100\%$

$$2 = \frac{60 - n_r}{60} \times 100$$

i.e. $\quad n_r = 60 - \frac{2 \times 60}{100} = 58.8 \text{ rev/s}$

i.e. the rotor runs at 58.8 × 60 = **3528 rev/min**

(c) Since the synchronous speed is 60 rev/s and that of the rotor is 58.8 rev/s. the rotating magnetic field cuts the rotor bars at (60–58.8), i.e. 1.2 rev/s. **Thus the frequency of the e.m.f.'s induced in the rotor bars is 1.2 Hz.**

C. FURTHER PROBLEMS

SHORT ANSWER PROBLEMS

1 Name three advantages that a three-phase induction motor has when compared with a d.c. motor.
2 Name the principal disadvantage of a three-phase induction motor when compared with a d.c. motor.
3 Write down the two properties of the magnetic field produced by the stator of a three-phase induction motor.
4 The speed at which the magnetic field of a three-phase induction motor rotates is called the .... speed.
5 The synchronous speed of a three-phase induction motor is .... proportional to the supply frequency.
6 The synchronous speed of a three-phase induction motor is .... proportional to the number of pairs of poles.
7 The type of rotor most widely used in a three-phase induction motor is called a .... rotor.
8 The slip of a three-phase induction motor is given by:

$$s = \frac{\ldots\ldots\ldots\ldots\ldots}{\ldots\ldots\ldots\ldots\ldots} \times 100\%.$$

9 A typical value for the slip of a small three-phase induction motor at full load is .... %.
10 As the load on the rotor of a three-phase induction motor increases, the slip ....

MULTI-CHOICE PROBLEMS (answers on page 191)

1 Which of the following statements about a three-phase squirrel-cage induction motor is false?
   (a) It has no external electrical connections to its rotor.
   (b) A three-phase supply is connected to its stator.
   (c) A magnetic flux which alternates is produced.
   (d) It is cheap, robust and requires little or no skilled maintenance.
2 Which of the following statements about a three-phase induction motor is false?
   (a) The speed of rotation of the magnetic field is called the synchronous speed.
   (b) A three-phase supply connected to the rotor produces a rotating magnetic field.
   (c) The rotating magnetic field has a constant speed and constant magnitude.
   (d) It is essentially a constant speed type machine.

3 Which of the following statements is false when referring to a three-phase induction motor?
(a) The synchronous speed is half the supply frequency when it has 4 poles.
(b) In a two-pole machine, the synchronous speed is equal to the supply frequency.
(c) If the number of poles is increased, the synchronous speed is reduced.
(d) The synchronous speed is inversely proportional to the number of poles.

4 A 4-pole, three-phase induction motor has a synchronous speed of 25 rev/s. The frequency of the supply to the stator is:
(a) 50 Hz; (b) 100 Hz; (c) 25 Hz; (d) 12.5 Hz.

Problems 5 and 6 refer to a three-phase induction motor. Which statements are false?

5 (a) The slip speed is the synchronous speed minus the rotor speed.
(b) As the rotor is loaded, the slip decreases.
(c) The frequency of induced rotor e.m.f.'s increases with load on the rotor.
(d) The torque on the rotor is due to the interaction of magnetic fields.

6 (a) If the rotor is running at synchronous speed, there is no torque on the rotor.
(b) If the number of poles on the stator is doubled, the synchronous speed is halved.
(c) At no load, the rotor speed is very nearly equal to the synchronous speed.
(d) The direction of rotation of the rotor is opposite to the direction of rotation of the magnetic field to give maximum current induced in the rotor bars.

A three-phase, 4-pole, 50 Hz induction motor runs at 1440 rev/min. In *Problems 7 to 10*, determine the correct answers for the quantities stated, selecting your answer from the list given.

(a) 12½ rev/s; (b) 25 rev/s; (c) 1 rev/s; (d) 50 rev/s;
(e) 1%; (f) 4%; (g) 50%; (h) 4 Hz;
(i) 50 Hz; (j) 1 Hz.

7 The synchronous speed.
8 The slip speed.
9 The percentage slip.
10 The frequency of induced e.m.f.'s in the rotor.

CONVENTIONAL PROBLEMS

1 With the aid of diagrams, explain how a rotating magnetic field is produced when a three-phase supply is connected to the stator of an induction motor.
2 Explain briefly what you understand by the term 'a six-pole induction motor'.
3 The synchronous speed of a 3-phase, 4-pole induction motor is 60 rev/s. Determine the frequency of the supply to the stator windings. [120 Hz]
4 The synchronous speed of a 3-phase induction motor is 25 rev/s and the frequency of the supply to the stator is 50 Hz. Calculate the equivalent number of pairs of poles of the motor. [2]
5 A 6-pole, 3-phase induction motor is connected to a 300 Hz supply. Determine the speed of rotation of the magnetic field produced by the stator. [100 rev/s]
6 A 6-pole, 3-phase induction motor runs at 970 rev/min at a certain load. If the stator is connected to a 50 Hz supply, find the percentage slip at this load. [3%]

7 A 3-phase, 50 Hz induction motor has 8 poles. If the full load slip is 2½%, determine (a) the synchronous speed, (b) the rotor speed and (c) the frequency of the rotor e.m.f.'s. [(a) 750 rev/min; (b) 731 rev/min; (c) 1.25 Hz]

8 Explain briefly, with the aid of sketches, the principle of operation of a 3-phase induction motor.

9 Explain how slip-frequency currents are set up in the rotor bars of a 3-phase induction motor and why this frequency varies with load.

10 Explain why a 3-phase induction motor develops no torque when running at synchronous speed. Define the slip of an induction motor and explain why its value depends on the load on the rotor.

11 A 4-pole, 3-phase, 50 Hz induction motor runs at 1440 rev/min at full load. Calculate (a) the synchronous speed, (b) the slip and (c) the frequency of the rotor induced e.m.f.'s. [(a) 1500 rev/min; (b) 4%; (c) 2 Hz]

12 A 12-pole, 3-phase, 50 Hz induction motor runs at 475 rev/min. Determine (a) the slip speed, (b) the percentage slip and (c) the frequency of rotor currents. [(a) 25 rev/min; (b) 5%; (c) 2.5 Hz]

8 Modulation

A. MAIN POINTS CONCERNED WITH MODULATION

1 The transmission of information such as speech, music and data over long distances requires the use of a carrier channel. It is common practice to 'carry' different communications, often called **signals**, at different frequencies to stop one signal from interfering with another. A signal can be shifted bodily from its original frequency band to another, this being achieved by 'modulating' one waveform with another.

The mean frequency level to which a signal is moved is called the **carrier frequency** and the process of superimposing the information signal on the carrier is called **modulation**. The resultant signal is called the **modulated signal**. Many signals, such as telephone conversations, can be transmitted simultaneously along a single pair of lines by using modulation techniques. Modulation of a band of low frequencies onto a higher frequency carrier is fundamental to radio communications and using different carrier frequencies leads to numerous programmes being transmitted simultaneously. The carrier frequency is the frequency to which the receiver has to be tuned, for example, about 88 to 90 MHz for Radio 2, the signal which is heard being obtained from the modulated carrier by a process called **demodulation**.

2 The carrier frequency must have one or more of its characteristics (i.e. amplitude, frequency and/or phase), varied by the information signal. When the amplitude of the carrier is changed by the information signal, the process is called **amplitude modulation**. To illustrate amplitude modulation, consider the signal to be a sinewave of frequency f_m, as shown in *Fig 1(a)* and the carrier to be a sinewave of frequency f_c, as shown in *Fig 1(b)*. The result of amplitude modulation is shown in *Fig 1(c)*, the signal information being duplicated on both sides of the carrier, as shown by the broken lines, which are construction lines outlining the pattern of change of amplitude of modulated waveform. This results in a band of frequencies over a range $(f_c - f_m)$ to $(f_c + f_m)$, i.e. the carrier frequency $\pm$ the signal frequency band. The frequency range between the highest and lowest of these frequencies is called the **bandwidth** (see *Problem 1*).

3 Instead of varying the amplitude of the carrier waveform, the modulating signal may be used to vary the frequency of the carrier. An increase in signal amplitude then causes a change in the modulated signal frequency, which is proportional to the amplitude of the modulating signal. This is called **frequency modulation** and is shown for a cosine wave signal in *Fig 2*.

When the signal amplitude is positive, the frequency of the carrier is modulated to be less than it was originally, shown as (a). The original carrier is

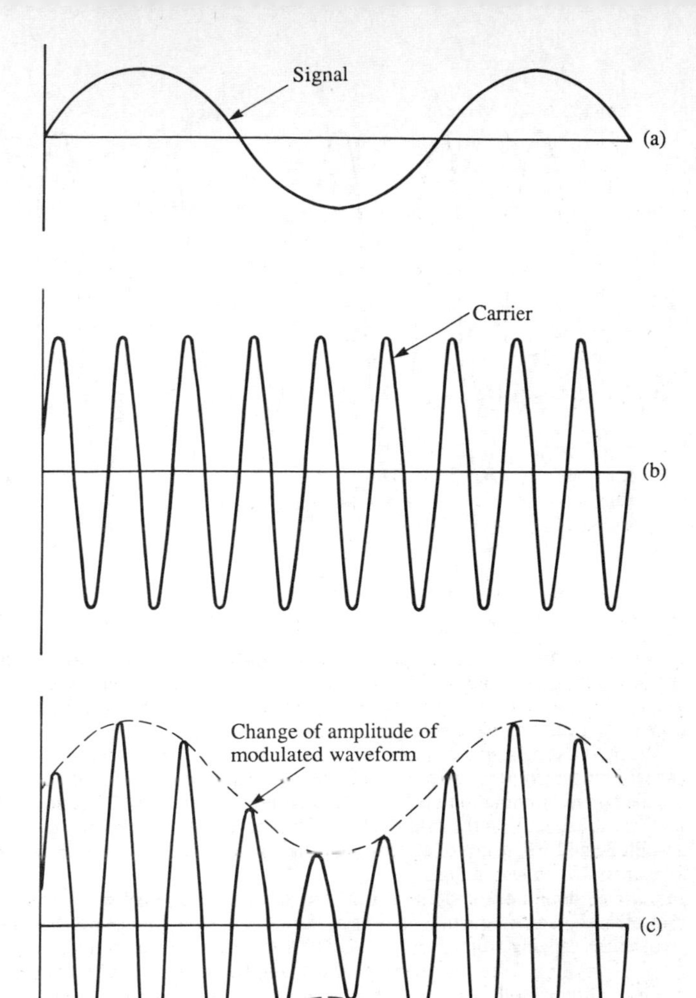
Signal
(a)
Carrier
(b)
Change of amplitude of
modulated waveform
(c)

Fig 1

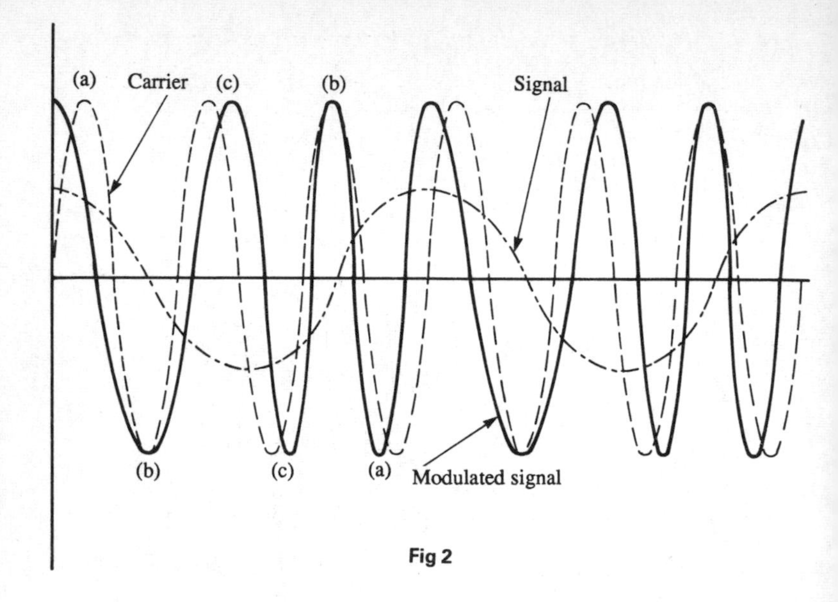

Fig 2

shown for reference. The modulated wave is in the same position as the original carrier when the signal amplitude is zero, as shown at (b). When the signal amplitude is negative, the frequency of the carrier is modulated to be greater than that of the original carrier, as shown at (c).

The modulating signal can be used to advance or retard the phase of the carrier in proportion to the amplitude of the modulating signal. This technique is called **phase modulation** and this also involves a variation of frequency. In this case it depends on the rate of change of phase and thus on both the amplitude and frequency of the modulating signal. The waveform shown is similar to that shown in *Fig 2*.

4 In **pulse modulation**, the signal is sampled at a frequency which is at least twice that of the highest frequency present in the signal. Thus for speech, having frequencies ranging from about 300 Hz to 3.4 kHz, a typical sampling frequency is 8 kHz. Various forms of pulse modulation are used and include pulse amplitude modulation, pulse position modulation and pulse duration modulation (see *Problem 5*).

In pulse code modulation, the signal amplitude is divided into a number of equal increments, each increment or level being designated by a number. For example, an amplitude divided into eight increments can have the instantaneous value of the amplitude transmitted by using the natural binary numbering system and three bits, the levels being transmitted as: 000, 001, 010, 011, ..., 111. This concept is shown in *Fig 3*.

Thus in pulse code modulation, an analogue signal is converted into a digital signal. Since the analogue signal can have any value between certain limits, but the resulting digital signal has only discrete values, some distortion of the signal results. The greater the number of increments used, the more closely the digital signal resembles the analogue signal.

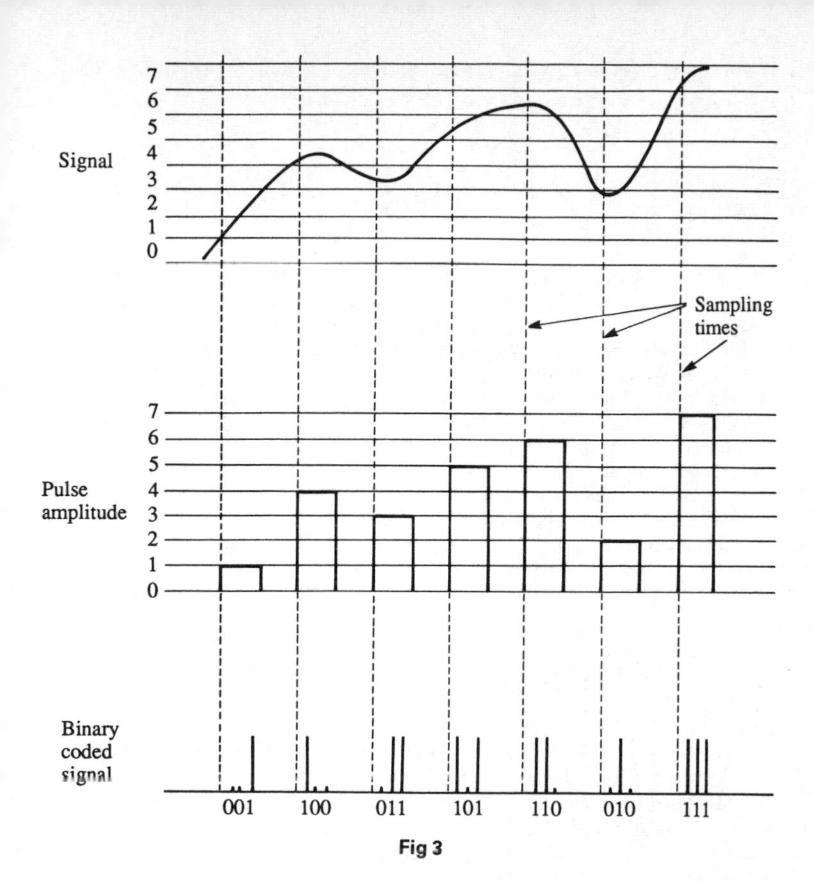

Fig 3

B. WORKED PROBLEMS ON MODULATION

Problem 1 A carrier of frequency 0.8 MHz is amplitude modulated by a signal having frequency components of 2 kHz, 4 kHz and 7 kHz. Determine the bandwidth of the modulated signal.

The highest frequency is 800 000 + 7000 = 807 000 Hz, i.e. 807 kHz.
The lowest frequency is 800 000 – 7000 = 793 000 Hz, i.e. 793 kHz.
From *para 2*, bandwidth is the frequency range between the highest and lowest frequencies thus:

bandwidth = 807 – 793 kHz
= 14 kHz

The bandwidth is also given by twice the highest modulating frequency. Thus,

$$\text{bandwidth} = 2 \times 7 \text{ kHz}$$
$$= 14 \text{ kHz}$$

Problem 2 Explain the term demodulation and state briefly how it is achieved.

When transmitting a signal over long distances, the process of superimposing the information signal on a carrier is called modulation. To extract the information signal at the receiving end, the process of modulation has to be reversed and is called **demodulation**. Consider a carrier having a frequency of 10 kHz on which an information signal of frequency range 4 kHz has been superimposed. The frequency range of the modulated signal is from 6 kHz to 14 kHz. When several carriers of different frequencies are being used, the modulated waveform must be isolated from the other carriers and their information signals. This is achieved by using a band pass filter (see chapter 10). Since the original signal has a range of 4 kHz and occupies ±4 kHz, that is 8 kHz as a modulated signal, it is necessary to reduce it to a signal of the original frequency range. This can be achieved by a rectifier circuit which removes the negative half of the modulated wave, as shown in *Fig 4*. However, most demodulation systems are more complex than the circuit shown.

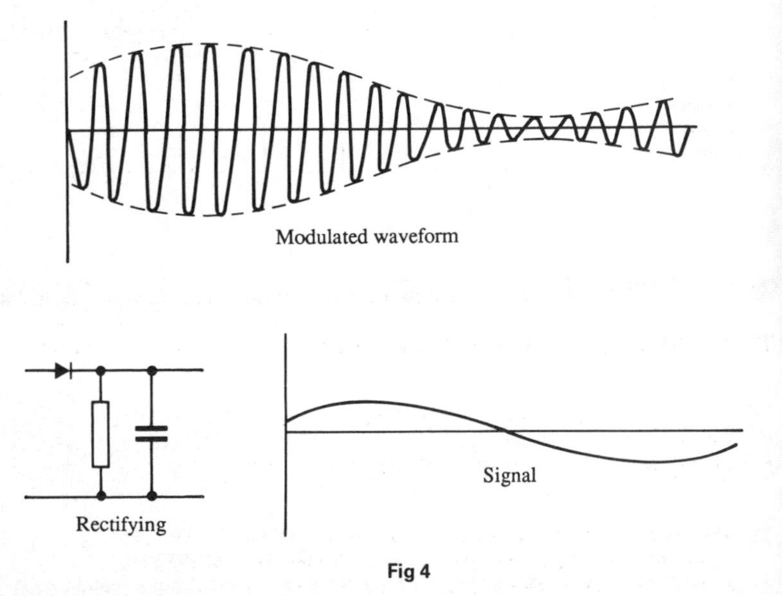

Fig 4

Problem 3 Define the terms frequency deviation, frequency swing and modulation index, when used in connection with frequency modulation.

Frequency deviation is a term used in frequency modulation and is defined as the peak difference between the instantaneous frequency of the modulated wave and the carrier frequency during one cycle of modulation.

Frequency swing is the difference between the maximum and minimum values of the instantaneous frequency of a frequency modulated wave.

The **modulating index** for a sinusoidal modulating waveform is the ratio of frequency deviation to the frequency of the modulating wave. Thus the modulating index is the ratio of the frequency deviation caused by a particular signal to the frequency of that signal.

Problem 4 Compare amplitude modulation and frequency modulation in relation to bandwidth, power and signal-to-noise ratio.

The principal advantage of amplitude modulation is the relatively simple circuitry used at the transmitting and receiving ends of the system. Such a system is used for medium and long wave sound broadcasting. The amplitude of the carrier is made to follow the fluctuations of the modulating signal. When the modulating signal occupies a bandwidth of B, the resulting bandwidth of the modulated signal is $2B$ and it has two symmetrical sidebands. Since either of the sidebands contains all the information, this method of transmission is wasteful of both bandwidth and power. In amplitude modulation much of the power is in the carrier, which does not contain any information. However, techniques are available which can eliminate one of the sidebands and when used, amplitude modulation is more economical of bandwidth when compared with frequency modulation. Frequency modulation uses less power than amplitude modulation due to power being transferred from the carrier to the side frequencies. Also, since the power being used in frequency modulation is constant, the transmitter output stages are designed to operate at their maximum efficiency all the time. The ratio of the required signal to the unrequired signal is called the **signal-to-noise ratio** and is measured in decibels. (See *para 1*, chapter 9.) Thus signal-to-noise ratio (dB) = 10 lg [(power in signal)/(power in noise)]. The effect of a noise signal mixing with a carrier is to produce a variation of amplitude, usually a small variation of phase and a consequent small variation in the resultant frequency. The variation of amplitude is the predominant effect and results in amplitude modulation having a higher signal-to-noise ratio than frequency modulation. In most frequency modulated systems the signal frequency variations swamp the much smaller frequency changes due to the interference.

Problem 5 With the aid of a sketch, briefly explain the terms pulse amplitude modulation, pulse duration modulation and pulse position modulation.

The principle of pulse amplitude modulation is shown in *Fig 5(a)*, in which the amplitude of the pulse is proportional to the amplitude of the signal. The amplitude of the pulse may change during the 'on' period or alternatively it may be kept constant, resulting in the stepped waveform as shown in *Fig 3*.

Fig 5(b) shows the principle of pulse duration modulation, the duration of the pulse being proportional to the amplitude of the signal. The position of the

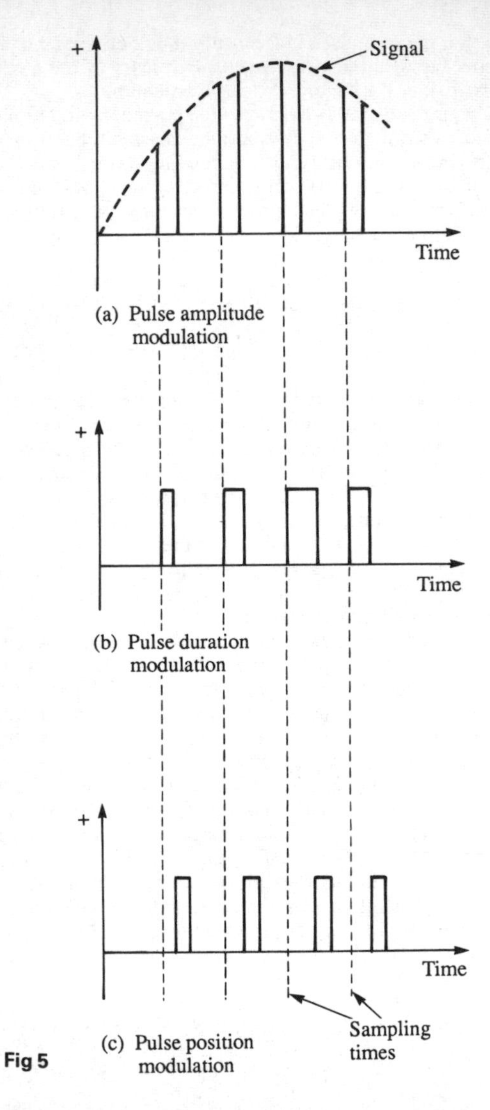

Fig 5

pulse relative to some datum (such as the sampling time), is made proportional to the amplitude of the signal in pulse position modulation, as shown in *Fig 5(c)*.

Problem 6 Explain the term **modulation index** as applied to amplitude modulation and with the aid of a sketch show the effect of overmodulation.

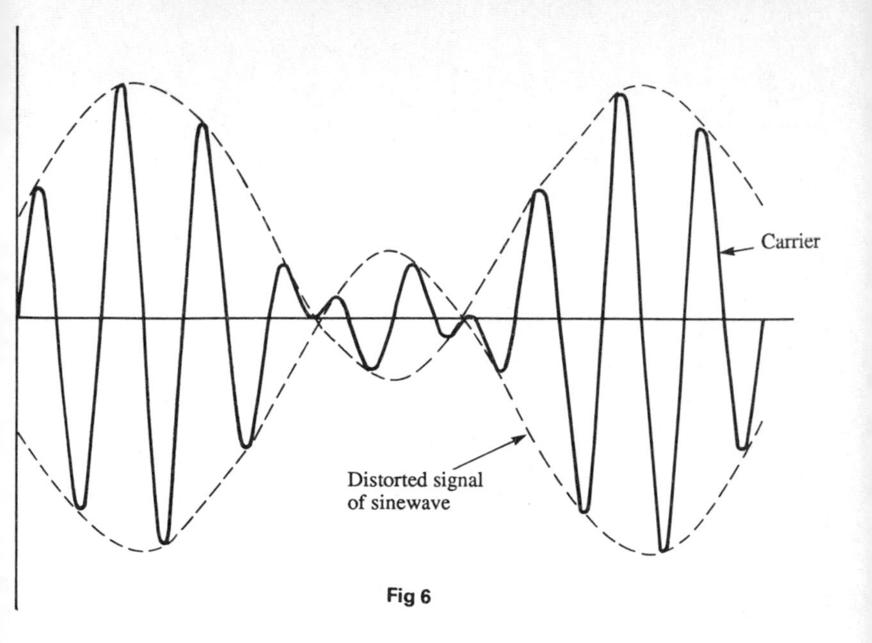

Fig 6

In amplitude modulation, let a carrier be a sinewave of the form

$a = A \sin \omega_c t = \phi$

When this is modulated by a sinewave of the form $b = B \sin \omega_m t$, the ratio B/A is called the modulating index. (Other terms such as modulating factor are used).

When the modulating index exceeds unity, distortion of the signal occurs, as is shown in *Fig 6*.

This shows that the modulating index of an amplitude modulated wave should not exceed unity. In amplitude modulation, the amplitude of the carrier is constant thus the modulating index is proportional to the amplitude of the modulating signal.

Problem 7 A carrier frequency of 2182 kHz is amplitude modulated by a telephony signal which contains frequencies ranging between 200 Hz and 3 kHz. Determine (a) the limits of the sidebands, (b) the overall bandwidth and (c) the frequency gap between the sidebands.

A modulated signal is invariably a complex waveform that is not sinusoidal and containing a range of frequencies. Let f_1 and f_2 be the lowest and highest frequencies of the modulating signal and let f_c be the carrier frequency. The lower sideband ranges from $(f_c - f_1)$ to $(f_c - f_2)$. The upper band of side frequencies ranges from $(f_c + f_2)$ to $(f_c + f_1)$, as shown in *Fig 7*.

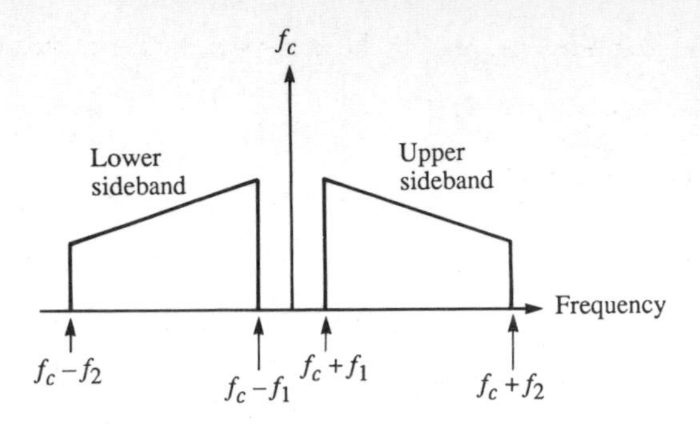

Fig 7

The frequencies corresponding to f_1, f_2 and f_c in this problem are:
$f_1 = 200$ Hz, $f_2 = 3$ kHz and $f_c = 2182$ kHz.

(a) The limits of the lower sideband are from (21821 – 0.2) to (2182 – 3) kHz, that is from 2181.8 kHz to 2179 kHz.
The limits of the upper sideband are from (2182 + 3) to (2182 + 0.2) kHz, that is from 2183 kHz to 2182.2 kHz.

(b) The bandwidth is from $(f_c + f_2)$ to $(f_c - f_2)$. But $(f_c + f_2) - (f_c - f_2) = 2 \times f_2 = 6$ kHz.

(c) The frequency gap between the sidebands is from $(f_c + f_1)$ to $(f_c - f_1)$. But $(f_c + f_1) - (f_c - f_1) = 2 \times f_1 = 2 \times 200 = 400$ Hz.

C. FURTHER PROBLEMS ON MODULATION

SHORT ANSWER PROBLEMS

1 What is meant by the term carrier when used in connection with modulation?
2 State an advantage of modulating a signal.
3 Name three characteristics of a carrier wave which can be changed when it is modulated.
4 Define bandwidth in terms of the carrier and signal frequency bands.
5 Sketch two cycles of an amplitude modulated wave.
6 Explain how frequency modulation differs from amplitude modulation.
7 Briefly explain the basic principles of pulse modulation.
8 Explain why some of the information signal is lost during pulse modulation.
9 Explain briefly why demodulation is necessary.
10 State where a filter may be used in a modualtion system.

MULTI-CHOICE PROBLEMS (answers on page 191)

Select the correct answer in the following problems:

1 In amplitude modulation
(a) the amplitude and phase of the carrier varies at the modulating frequency

(b) the amplitude and frequency of the carrier varies at the modulating frequency
(c) the bandwidth is given by $(f_c - f_m)$
(d) the carrier must have a higher frequency than the information signal.

2 In frequency modulation:
(a) the frequency of the carrier is varied by the amplitude of the modulating signal
(b) the modulating signal is used to alter the phase of the carrier in proportion to the amplitude of the modulating signal
(c) the amplitude of the carrier varies in proportion to the amplitude of the modulating signal
(d) the signal causes the frequency of the modulating signal to vary.

3 In pulse modulation:
(a) the amplitude, duration or phase of the pulse may be varied
(b) the information is transmitted as a continuous signal
(c) the signal is sampled at a frequency which is at least twice the frequency of the carrier
(d) the transmitted signal is an analogue signal.

4 A carrier of frequency 1 MHz is amplitude modulated by a signal having a largest frequency component of 10 kHz. The bandwidth of the modulated signal is:
(a) 0.9 MHz (b) 20 kHz (c) 10 kHz (d) 1.1 MHz.

5 Compared with frequency modulation, amplitude modulation:
(a) has a better signal-to-noise ratio
(b) uses less power
(c) uses more expensive equipment in transmitting and receiving
(d) can have a smaller bandwidth.

CONVENTIONAL PROBLEMS

1 Explain why modulation is frequently used when transmitting a signal over long distances, with reference to (a) a telephone and (b) a broadcasting system.

2 Sketch two cycles of an amplitude modulated waveform in which the carrier has a frequency of ten times the modulating wave and having a modulation factor of 0.4.

3 A carrier of frequency 0.7 MHz is amplitude modulated by a signal, the frequency components of which are 1 kHz, 2 kHz and 5 kHz. Determine:
(a) the bandwidth of the resultant modulated signal and
(b) the frequency range covered by the lower sideband.
List the side frequencies.
[(a) 10 kHz, (b) 695 to 699 kHz, 695, 698, 699, 702, 201 and 705 kHz]

4 A carrier of frequency 800 kHz is amplitude modulated by a signal containing frequencies ranging from 50 Hz to 12 kHz. Determine:
(a) the frequency limits of the sidebands,
(b) the overall bandwidth of the modulated wave and
(c) the frequency gap between the sidebands.
[(a) 788 to 799.95 kHz and 800.05 to 812 kHz, (b) 24 kHz, (c) 100 Hz]

5 With the aid of a sketch describe a system which converts an analogue signal system into a digital signal system.

6 Discuss the advantages and disadvantages of an amplitude modulated system when compared with a frequency modulated system.

9 Measuring instruments and measurements

A. MAIN POINTS CONCERNED WITH MEASURING INSTRUMENTS AND MEASUREMENTS

1 In electronic systems, the ratio of two similar quantities measured at different points in the system, are often expressed in logarithmic units. By definition, if the ratio of two powers P_1 and P_2 is to be expressed in **decibel (dB) units** then the number of decibels, X, is given by:

$$\mathbf{X = 10\ lg\left(\frac{P_2}{P_1}\right)\ dB} \tag{1}$$

Thus, when the power ratio, $\frac{P_2}{P_1} = 1$ then the decibel power ratio $= 10\ lg\ 1 = 0$

when the power ratio, $\frac{P_2}{P_1} = 100$ then the decibel power ratio $= 10\ lg\ 100 = +20$

(i.e. a power gain),

and when the power ratio, $\frac{P_2}{P_1} = \frac{1}{10}$ then the decibel power ratio $= 10\ lg\ \frac{1}{10} = -10$,

(i.e. a power loss or attenuation).

2 Logarithmic units may also be used for voltage and current ratios. Power, P, is given by $P = I^2 R$ or $P = V^2/R$.
Substituting in equation (1) gives:

$$X = 10\ lg\left(\frac{I_2^{\,2} R_2}{I_1^{\,2} R_1}\right)\ dB \quad \text{or} \quad X = 10\ lg\left(\frac{V_2^{\,2}/R_2}{V_1^{\,2}/R_1}\right)\ dB$$

If $R_1 = R_2$, then $X = 10\ lg\left(\frac{I_2^{\,2}}{I_1^{\,2}}\right)$ dB or $X = 10\ lg\left(\frac{V_2^{\,2}}{V_1^{\,2}}\right)$ dB

i.e. $$\mathbf{X = 20\ lg\left(\frac{I_2}{I_1}\right)\ db \quad or \quad X = 20\ lg\left(\frac{V_2}{V_1}\right)\ dB}$$

(from the laws of logarithms).

3 From equation (1), X decibels is a logarithmic ratio of two similar quantities and is not an absolute unit of measurement. It is therefore necessary to state a **reference level** to measure a number of decibels above or below that reference. The most widely used reference level for power is 1 mW, and when power levels

are expressed in decibels, above or below the 1 mW reference level, the unit given to the new power level is dBm.

A voltmeter can be re-scaled to indicate the power level directly in decibels. The scale is generally calibrated by taking a reference level of 0 dB when a power of 1 mW is dissipated in a 600 Ω resistor (this being the natural impedance of a simple transmission line). The reference voltage V is then obtained from

$P = \frac{V^2}{R}$, i.e. $1 \times 10^{-3} = \frac{V^2}{600}$ from which, $V = 0.775$ volts.

In general, the number of dBm, $X = 20 \lg \left(\frac{V}{0.775}\right)$.

Thus $V = 0.20$ V corresponds to $20 \lg \left(\frac{0.20}{0.775}\right) = -11.77$ dBm and

$V = 0.90$ V corresponds to $20 \lg \left(\frac{0.90}{0.775}\right) = +1.3$ dBm, and so on.

A typical decibelmeter, or dB meter, scale is shown in *Fig 1*. Errors are introduced with dB meters when the circuit impedance is not 600 Ω. (See *Problems 1 to 6*).

4 The **cathode ray oscilloscope (CRO)** may be used for the observation of waveforms and for the measurement of voltage, current, frequency, phase and periodic time. Double beam oscilloscopes are useful whenever two signals are to be compared simultaneously. The CRO demands reasonable skill in adjustment and use. However its greatest advantage is in observing the shape of a waveform – a feature not possessed by other measuring instruments (see *Problems 7 to 10*).

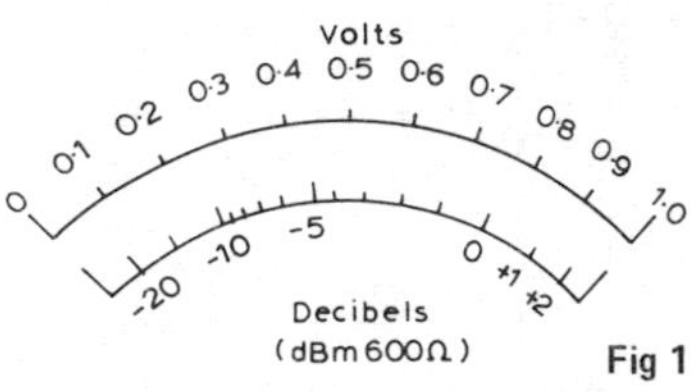

Fig 1

5 A **Wheatstone bridge**, shown in *Fig 2(a)* is used in d.c. circuits to compare an unknown resistance R_x with others of known values. R_3 is varied until zero deflection is obtained on the galvanometer, G. At balance (i.e. zero deflection on

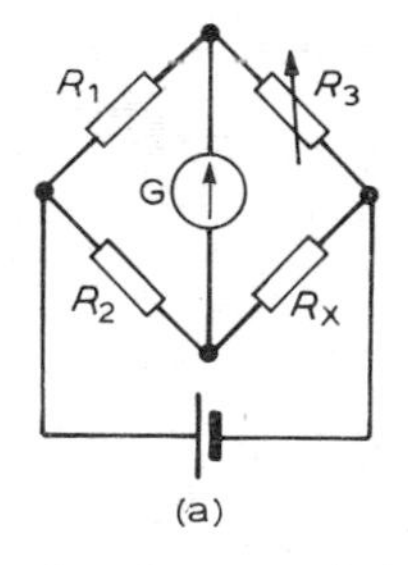

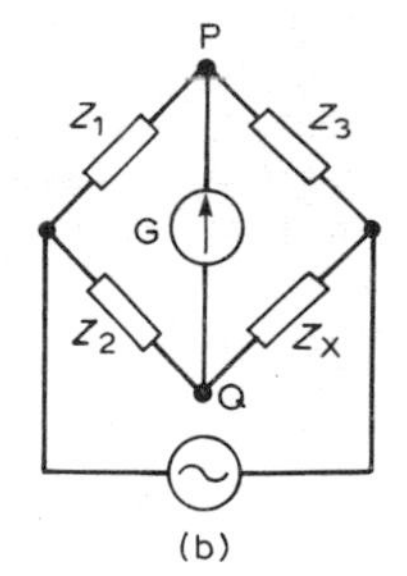

Fig 2

the galvanometer) the products of diagonally opposite resistances are equal to one another,

i.e. $R_1 R_x = R_2 R_3$

from which $R_x = \frac{R_2 R_3}{R_1}$ **ohms.**

6 A Wheatstone bridge type circuit, shown in *Fig 2(b)*, may be used in a.c. circuits to determine unknown values of inductance and capacitance, as well as resistance.

When the potential differences across Z_3 and Z_x (or across Z_1 and Z_2) are equal in magnitude and phase, then the current flowing through the galvanometer, G, is zero.

At balance, $Z_1 Z_x = Z_2 Z_3$, from which, $Z_x = \dfrac{Z_2 Z_3}{Z_1}$ Ω.

7 There are many forms of a.c. bridge, and these include: the Maxwell, Hay, Owen and Heaviside bridges for measuring inductance, and the De Sauty, Schering and Wien bridges for measuring capacitance. A **commercial or universal bridge** is one which can be used to measure resistance, inductance or capacitance.

8 Maxwell's bridge for measuring the inductance L and resistance r of an inductor is shown in *Fig 3*.

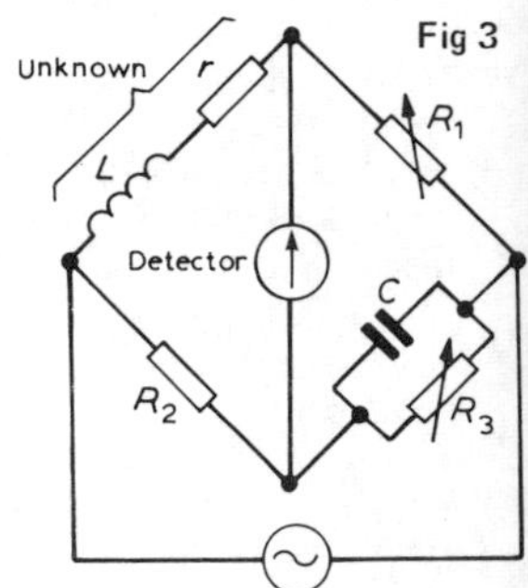

At balance: $L = R_1 R_2 C$ **henrys** (1)

and $r = \dfrac{R_1 R_2}{R_3}$ **ohms** (2)

(see *Problems 11 and 12*).

From equation (1), $R_2 = \dfrac{L}{R_1 C}$ and from

equation (2), $R_3 = \dfrac{R_1}{r} R_2$.

Hence $R_3 = \dfrac{R_1}{r}\left(\dfrac{L}{R_1 C}\right) = \dfrac{L}{Cr}$

If the frequency is constant then $R_3 \propto \dfrac{L}{r} \propto \dfrac{\omega L}{r} \propto$ Q-factor.

Thus the bridge can be adjusted to give a direct indication of Q-factor.

9 The **Q-factor** for a series L–C–R circuit is the voltage magnification at resonance,

i.e. Q-factor $= \dfrac{\text{voltage across capacitor}}{\text{supply voltage}} = \dfrac{V_c}{V}$ (see chapter 2, para. 15).

The simplified circuit of a **Q-meter**, used for measuring Q-factor, is shown in *Fig 4*. Current from a variable frequency oscillator flowing through a very low resistance r develops a variable frequency voltage, V_r, which is applied to a series L–R–C circuit. The frequency is then varied until resonance causes voltage V_c to reach a maximum value. At resonance V_r and V_c are noted.

Then Q-factor $= \dfrac{V_c}{V_r} = \dfrac{V_c}{Ir}$

In a practical Q-meter, V_r is maintained constant and the electronic voltmeter can be calibrated to indicate the Q-factor directly. If a variable capacitor C is used and the oscillator is set to a given frequency, then C can be adjusted to give resonance. In this way inductance L may be calculated using

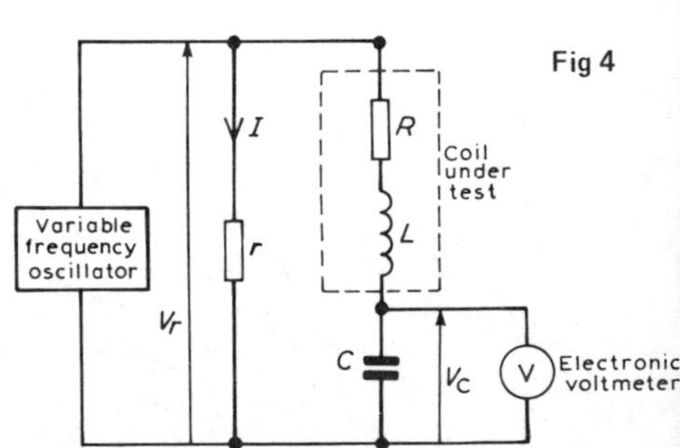

$$f_r = \frac{1}{2\pi\sqrt{(LC)}}.$$

Since $Q = \frac{2\pi f L}{R}$, then R may be calculated.

Q-meters operate at various frequencies and instruments exist with frequency ranges from 1 kHz to 50 MHz. Errors in measurement can exist with Q-meters since the coil has an effective parallel self capacitance due to capacitance between turns. The accuracy of a Q-meter is approximately ± 5% (see *Problem 13*).

10 **Waveform harmonics**

(i) Let an instantaneous voltage v be represented by $v = V_M \sin 2\pi ft$ volts. This is a waveform which varies sinusoidally with time t, has a frequency f, and a maximum value V_M. Alternating voltages are usually assumed to have wave-shapes which are sinusoidal where only one frequency is present. If the wave-form is not sinusoidal it is called a **complex wave**, and, whatever its shape, it may be split up mathematically into components called the **fundamental** and a number of **harmonics.** This process is called harmonic analysis. The fundamental (or first harmonic) is sinusoidal and has the supply frequency, f; the other harmonics are also sine waves having frequencies which are integer multiples of f. Thus, if the supply frequency is 50 Hz, then the third harmonic frequency is 150 Hz, the fifth 250 Hz, and so on.

(ii) A complex waveform comprising the sum of the fundamental and a third harmonic of about half the amplitude of the fundamental is shown in *Fig 5(a)*, both waveforms being initially in phase with each other. If further odd harmonic waveforms of the appropriate amplitudes are added, a good approximation to a square wave results. In *Fig 5(b)* the third harmonic is shown having an initial phase displacement from the fundamental. The positive and negative half cycles of each of the complex waveforms shown in *Figs 5(a) and (b)* are identical in shape, and this is a feature of waveforms containing the fundamental and only odd harmonics.

(iii) A complex waveform comprising the sum of the fundamental and a second harmonic of about half the amplitude of the fundamental is shown in *Fig 5(c)*, each waveform being initially in phase with each other. If further even harmonics of appropriate amplitudes are added a good approximation to a triangular wave results. In *Fig 5(c)* the negative cycle appears as a mirror image of the positive cycle about point A. In *Fig 5(d)* the second harmonic is shown with an initial phase displacement from the fundamental and the positive and negative half cycles are dissimilar.

(iv) A complex waveform comprising the sum of the fundamental, a second harmonic and a third harmonic is shown in *Fig 5(e)*, each waveform being initially 'in-phase'. The negative half cycle appears on a mirror image of the positive cycle about point B. In *Fig 5(f)*, a complex waveform comprising the sum of the fundamental, a second harmonic and a third harmonic are shown with initial phase displacement. The positive and negative half cycles are seen to be dissimilar.

The features mentioned relative to *Figs 5(a) to (f)* make it possible to recognise the harmonics present in a complex waveform displayed on a CRO.

11 Some measuring instruments depend for their operation on power taken from the circuit in which measurements are being made. Depending on the 'loading' effect of the instrument (i.e. the current taken to enable it to operate), the prevailing circuit conditions may change.

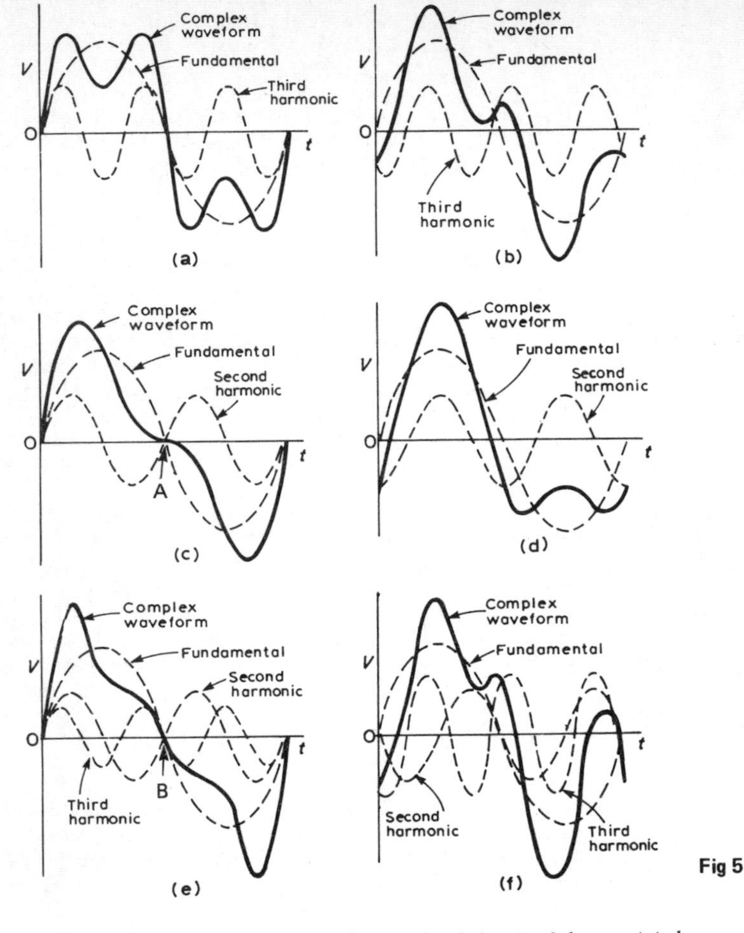

Fig 5

The resistance of voltmeters may be calculated since each have a stated sensitivity (or 'figure of merit'), often stated in 'kΩ per volt' of FSD. A voltmeter should have as high a resistance as possible (–ideally infinite). In a.c. circuits the impedance of the instrument varies with frequency and thus the loading effect of the instrument can change. (See *Problems 14 and 15*.)

12 Electronic measuring instruments have advantages over instruments such as the moving iron or moving coil meters, in that they have a much higher input resistance (some as high as 1000 MΩ) and can handle a much wider range of frequency (from d.c. up to MHz).

The digital voltmeter (DVM) is one which provides a digital display of the voltage being measured. Advantages of a DVM over analogue instruments include higher accuracy and resolution, no observational or parallax errors and a very high input resistance, constant on all ranges.

A digital multimeter is a DVM with additional circuitry which makes it capable of measuring a.c. voltage, d.c. and a.c. current and resistance.

Instruments for a.c. measurements are generally calibrated with a sinusoidal alternating waveform to indicate r.m.s. values when a sinusoidal signal is applied to the instrument. Some instruments, such as the moving iron and electro-dynamic instruments, give a true r.m.s. indication. With other instruments the indication is either scaled up from the mean value (such as with the rectifier moving coil instrument) or scaled down from the peak value.

Sometimes quantities to be measured have complex waveforms (see para. 10), and whenever a quantity is non-sinusoidal, errors in instrument readings can occur if the instrument has been calibrated for sine waves only. Such waveform errors can be largely eliminated by using electronic instruments.

13 Errors are always introduced when using instruments to measure electrical quantities. Besides possible errors introduced by the operator or by the instrument disturbing the circuit, errors are caused by the limitations of the instrument used.

The calibration accuracy of an instrument depends on the precision with which it is constructed. Every instrument has a margin of error which is expressed as a percentage of the instruments full scale deflection. For example, an instrument may have an accuracy of ± 2% of FSD. Thus, if a voltmeter has a FSD of 100 V and it indicates say 60 V, then the actual voltage measured may be anywhere between 60 V ± (2% of 100 V), i.e. 60 ± 2 V, i.e. between 58 V and 62 V.
As a percentage of the voltmeter reading this error is ± 2/60 × 100%, i.e. ± 3.33%. Hence the accuracy can be expressed as 60 V ± 3.33%. It follows that an instrument having a 2% FSD accuracy can give relatively large errors when operating at conditions well below FSD.

When more than one instrument is used in a circuit then a cumulative error results. For example, if the current flowing through and the p.d. across a resistor is measured, then the percentage error in the ammeter is added to the percentage error in the voltmeter when determining the maximum possible error in the measured value of resistance. (See *Problems 16 to 18.*)

B. WORKED PROBLEMS ON MEASURING INSTRUMENTS AND MEASUREMENTS

Problem 1 The ratio of two powers is (a) 2, (b) 30, (c) 1000, (d) $\frac{1}{20}$. Determine the decibel power ratio in each case.

From para. 1, the power ratio in decibels, X, is given by: $X = 10 \lg \left(\frac{P_2}{P_1}\right)$

(a) When $\frac{P_2}{P_1} = 2$, $X = 10 \lg (2) = 10(0.30) = \mathbf{3\ dB}$

(b) When $\frac{P_2}{P_1} = 30$, $X = 10 \lg (30) = 10(1.48) = \mathbf{14.8\ dB}$

(c) When $\frac{P_2}{P_1} = 1000$, $X = 10 \lg (1000) = 10(3.0) = \mathbf{30\ dB}$

(d) When $\frac{P_2}{P_1} = \frac{1}{20} = 0.05$, $X = 10 \lg (0.05) = 10(-1.30) = $ **-13 dB**

(a), (b) and (c) represent power gains and (d) represents a power loss or attenuation.

Problem 2 The current input to a system is 5 mA and the current output is 20 mA. Find the decibel current ratio assuming the input and load resistances of the system are equal.

From para. 2, the decibel current ratio $= 20 \lg \left(\frac{I_2}{I_1}\right) = 20 \lg \left(\frac{20}{5}\right) = 20 \lg 4$

$= 20\,(0.60)$

$=$ **12 dB gain**

Problem 3 6% of the power supplied to a cable appears at the output terminals. Determine the power loss in decibels.

If P_1 = input power and P_2 = output power then $\frac{P_2}{P_1} = \frac{6}{100} = 0.06$

Decibel power ratio $= 10 \lg \left(\frac{P_2}{P_1}\right) = 10 \lg (0.06) = 10(-1.222)$

$= -12.22$ dB

Hence the decibel power loss is 12.22 dB

Problem 4 An amplifier has a gain of 14 dB. Its input power is 8 mW. Find its output power.

Decibel power ratio $= 10 \lg \left(\frac{P_2}{P_1}\right)$ where P_1 = input power = 8 mW, and P_2 = output power.

Hence $\quad 14 = 10 \lg \left(\frac{P_2}{P_1}\right)$,

$1.4 = \lg \left(\frac{P_2}{P_1}\right)$,

and $\quad 10^{1.4} = \frac{P_2}{P_1}$ from the definition of a logarithm

i.e. $\quad 25.12 = \frac{P_2}{P_1}$

Output power, $P_2 = 25.12\,P_1 = (25.12)(8) =$ **201 mW or 0.201 W**

Problem 5 Determine, in decibels, the ratio of output power to input power of a 3 stage communications system, the stages having gains of 12 dB, 15 dB and -8 dB. Find also the overall power gain.

The decibel ratio may be used to find the overall power ratio of a chain simply by adding the decibel power ratios together.
Hence the overall decibel power ratio $= 12+15-8 =$ **19 dB gain.**

$$19 = 10 \lg \left(\frac{P_2}{P_1}\right)$$

Hence $\quad 1.9 = \lg \left(\frac{P_2}{P_1}\right)$ and $10^{1.9} = \frac{P_2}{P_1} = 79.4$

Thus the overall power gain, $\frac{P_2}{P_1} = 79.4$

[For the first stage, $12 = 10 \lg \left(\frac{P_2}{P_1}\right)$, from which $\frac{P_2}{P_1} = 10^{1.2} = 15.85$.

Similarly for the second stage, $\frac{P_2}{P_1} = 31.62$ and for the third stage, $\frac{P_2}{P_1} = 0.158$

The overall power ratio is thus $15.85 \times 31.62 \times 0.158 = 79.2$]

Problem 6 The output voltage from an amplifier is 4 V. If the voltage gain is 27 dB, calculate the value of the input voltage assuming that the amplifier input resistance and load resistance are equal.

Voltage gain in decibels $= 27 = 20 \lg \left(\frac{V_2}{V_1}\right) = 20 \lg \left(\frac{4}{V_1}\right)$

Hence $\quad \frac{27}{20} = \lg \left(\frac{4}{V_1}\right)$

$$1.35 = \lg \left(\frac{4}{V_1}\right)$$

$10^{1.35} = \frac{4}{V_1}$, from which $V_1 = \frac{4}{10^{1.35}} = \frac{4}{22.39} = 0.179$ V.

Hence the input voltage V_1 is 0.179 V

Problem 7 For the CRO square voltage waveform shown in *Fig 6* determine (a) the periodic time; (b) the frequency and (c) the peak to peak voltage. The 'time/cm' (or time base control) switch is on 100 μs/cm and the 'volts/cm' (or signal amplitude control) switch is on 20 V/cm.

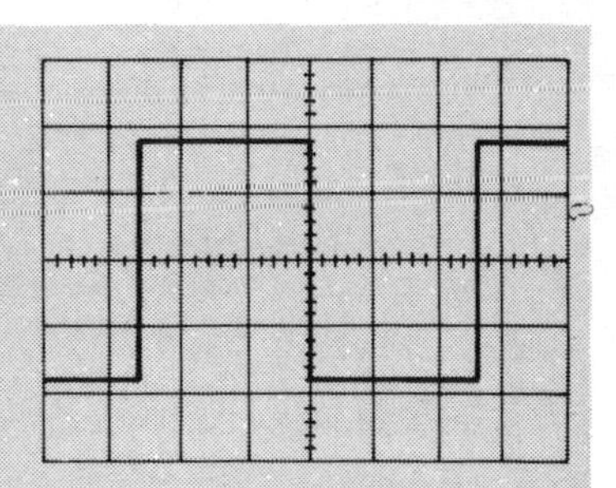

Fig 6

(In *Figs 6 to 9* assume that the squares shown are 1 cm by 1 cm.)

(a) The width of one complete cycle is 5.2 cm.
Hence the periodic time, $T = 5.2 \text{ cm} \times 100 \times 10^{-6}$ s/cm $=$ **0.52 ms**

(b) Frequency, $f = \frac{1}{T} = \frac{1}{0.52 \times 10^{-3}} =$ **1.92 kHz**

(c) The peak to peak height of the display is 3.6 cm
Hence the peak to peak voltage $= 3.6 \text{ cm} \times 20$ V/cm $=$ **72 V**

Problem 8 For the CRO display of a pulse waveform shown in *Fig 7* the 'time/cm' switch is on 50 ms/cm and the 'volts/cm' switch is on 0.2 V/cm. Determine (a) the periodic time; (b) the frequency and (c) the magnitude of the pulse voltage.

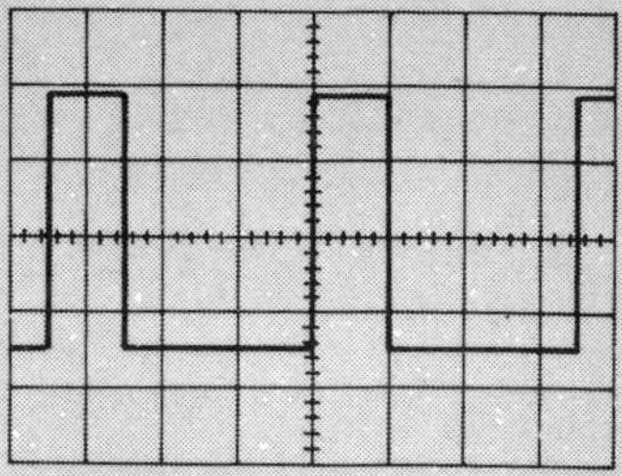

Fig 7

(a) The width of one complete cycle is 3.5 cm.
Hence the periodic time, $T = 3.5 \text{ cm} \times 50 \text{ ms/cm} = \mathbf{175 \text{ ms}}$

(b) Frequency, $f = \dfrac{1}{T} = \dfrac{1}{0.175} = \mathbf{5.71 \text{ Hz}}$

(c) The height of a pulse is 3.4 cm
Hence the magnitude of the pulse voltage $= 3.4 \text{ cm} \times 0.2 \text{ V/cm} = \mathbf{0.68 \text{ V}}$

Problem 9 A sinusoidal voltage trace displayed by a CRO is shown in *Fig 8*. If the 'time/cm' switch is on 500 μs/cm and the 'volts/cm' switch on 5 V/cm, find, for the waveform, (a) the frequency; (b) the peak to peak voltage (c) the amplitude and (d) the r.m.s. value.

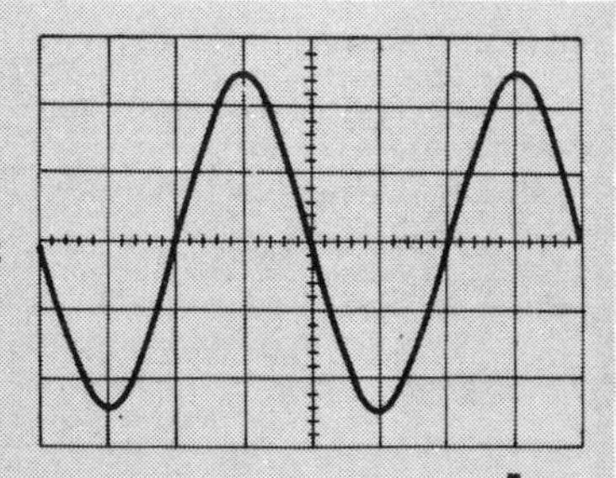

Fig 8

(a) The width of one complete cycle is 4 cm. Hence the periodic time, T, is 4 cm × 500 μs/cm, i.e. 2 ms.

Frequency, $f = \dfrac{1}{T} = \dfrac{1}{2 \times 10^{-3}} = \mathbf{500 \text{ Hz}}$

(b) The peak to peak height of the waveform is 5 cm.
Hence the peak to peak voltage $= 5 \text{ cm} \times 5 \text{ V/cm} = 25 \text{ V}$

(c) Amplitude $= \dfrac{1}{2} \times 25 \text{ V} = \mathbf{12.5 \text{ V}}$.

(d) The peak value of voltage is the amplitude, i.e. 12.5 V

r.m.s. voltage $= \dfrac{\text{peak voltage}}{\sqrt{2}} = \dfrac{12.5}{\sqrt{2}} = \mathbf{8.84 \text{ V}}$

Problem 10 For the double beam oscilloscope displays shown in *Fig 9* determine (a) their frequency, (b) their r.m.s. values and (c) their phase difference. The 'time/cm' switch is on 100 μs/cm and the 'volts/cm' switch on 2 V/cm.

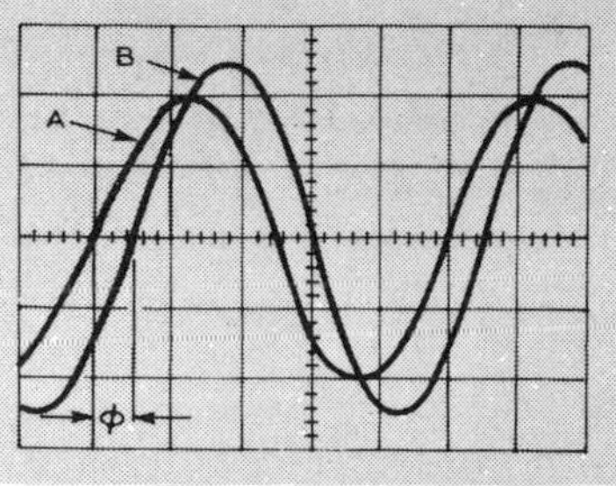

Fig 9

(a) The width of each complete cycle is 5 cm for both waveforms.
Hence the periodic time, T, of each waveform is 5 cm × 100 μs/cm, i.e. 0.5 ms

Frequency of each waveform, $f = \frac{1}{T} = \frac{1}{0.5 \times 10^{-3}} = \mathbf{2\ kHz}$

(b) The peak value of waveform A is 2 cm × 2 V/cm = 4 V

Hence the r.m.s. value of waveform A $= \frac{4}{\sqrt{2}} = \mathbf{2.83\ V}$

The peak value of waveform B is 2.5 cm × 2 V/cm = 5 V

Hence the r.m.s. value of waveform B $= \frac{5}{\sqrt{2}} = \mathbf{3.54\ V}$

(c) Since 5 cm represents 1 cycle, then 5 cm represents 360°, i.e. 1 cm represents $\frac{360}{5} = 72°$. The phase angle ϕ = 0.5 cm = 0.5 cm × 72°/cm = 36°

Hence waveform A leads waveform B by 36°

(An alternative method for finding frequency and phase angle is by Lissajous figures.)

Problem 11 Sketch a Maxwell bridge circuit arrangement suitable for measuring the inductance and resistance of a coil. Derive expressions for the inductance and resistance when the bridge is balanced.

A Maxwell bridge circuit is shown in *Fig 10*.
At balance the products of diagonally opposite impedances are equal.
Thus $Z_1 Z_2 = Z_3 Z_4$.

Using complex quantities, $Z_1 = R_1$, $Z_2 = R_2$,

$$Z_3 = \frac{R_3(-jX_C)}{R_3 - jX_C} \quad \text{i.e.} \ \frac{\text{product}}{\text{sum}}$$

and $Z_4 = r + jX_L$.

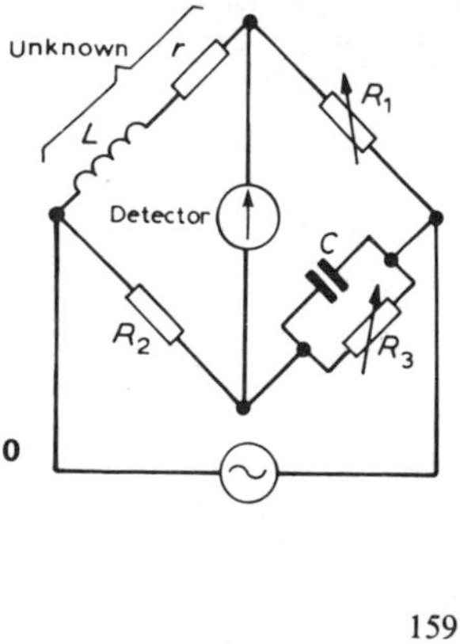

Fig 10

Hence $R_1 R_2 = \dfrac{R_3(-jX_C)}{(R_3-jX_C)}(r+jX_L)$

i.e. $R_1 R_2(R_3-jX_C) = (-jR_3 X_C)(r+jX_L)$

$R_1 R_2 R_3 - jR_1 R_2 X_C = -jrR_3 X_C - j^2 R_3 X_C X_L$

i.e. $R_1 R_2 R_3 - jR_1 R_2 X_C = -jrR_3 X_C + R_3 X_C X_L$ (since $j^2 = -1$)

Equating the real parts gives: $R_1 R_2 R_3 = R_3 X_C X_L$

from which $X_L = \dfrac{R_1 R_2}{X_C}$

i.e. $2\pi f L = \dfrac{R_1 R_2}{\dfrac{1}{2\pi f C}} = R_1 R_2 (2\pi f C)$

Hence inductance, $L = R_1 R_2 C$ henry (1)

Equating the imaginary parts gives: $-R_1 R_2 X_C = -rR_3 X_c$

from which, resistance, $r = \dfrac{R_1 R_2}{R_3}$ ohms (2)

Problem 12 For the a.c. bridge shown in *Fig 10* determine the values of the inductance and resistance of the coil when $R_1 = R_2 = 400\ \Omega$, $R_3 = 5\ \text{k}\Omega$ and $C = 7.5\ \mu\text{F}$.

From equation (1) above, inductance $L = R_1 R_2 C = (400)(400)(7.5 \times 10^{-6})$
$= \mathbf{1.2\ H}$.

From equation (2) above, resistance, $r = \dfrac{R_1 R_2}{R_3} = \dfrac{(400)(400)}{(5000)} = \mathbf{32\ \Omega}$

Problem 13 When connected to a Q-meter an inductor is made to resonate at 400 kHz. The Q-factor of the circuit is found to be 100 and the capacitance of the Q-meter capacitor is set to 400 pF. Determine (a) the inductance, and (b) the resistance of the inductor.

Resonant frequency, $f_r = 400\ \text{kHz} = 400 \times 10^3\ \text{Hz}$. Q-factor = 100;
Capacitance, $C = 400\ \text{pF} = 400 \times 10^{-12}\ \text{F}$.
The circuit diagram of a Q-meter is shown in *Fig 4*.

(a) At resonance, $f_r = \dfrac{1}{2\pi\sqrt{(LC)}}$ for a series L–C–R circuit.

Hence $2\pi f_r = \dfrac{1}{\sqrt{(LC)}}$, from which $(2\pi f_r)^2 = \dfrac{1}{LC}$

and inductance, $L = \dfrac{1}{(2\pi f_r)^2 C} = \dfrac{1}{(2\pi 400 \times 10^3)^2 (400 \times 10^{-12})}$ H

$= \mathbf{396\ \mu H\ or\ 0.396\ mH}$.

(b) Q-factor at resonance $= \dfrac{2\pi f_r L}{R}$

from which resistance $R = \dfrac{2\pi f_r L}{Q} = \dfrac{2\pi(400 \times 10^3)(0.396 \times 10^{-3})}{100} = \mathbf{9.95\ \Omega}$

Problem 14 A voltmeter having a FSD of 100 V and a sensitivity of 1.6 kΩ/V is used to measure voltage V_1 in the circuit of *Fig 11*. Determine (a) the value of voltage V_1 with the voltmeter not connected, and (b) the voltage indicated by the voltmeter when connected between A and B.

(a) By voltage division, $V_1 = \left(\frac{40}{40+60}\right) 100 = \mathbf{40\ V}$

(b) The resistance of a voltmeter having a 100 V FSD and sensitivity 1.6 kΩ/V is 100 V × 1.6 kΩ/V = 160 kΩ.

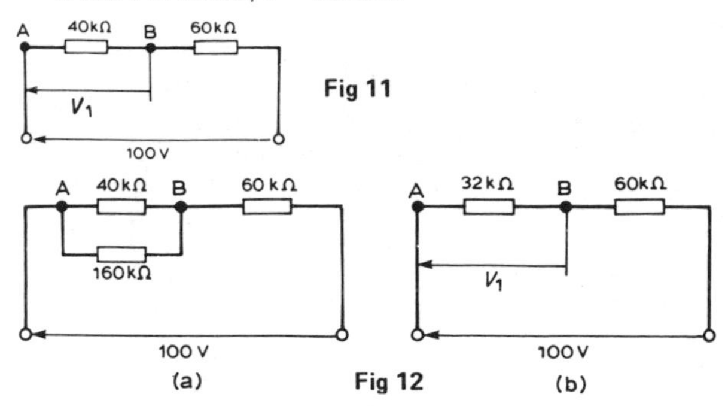

Fig 11

Fig 12

When the voltmeter is connected across the 40 kΩ resistor the circuit is as shown in *Fig 12(a)* and the equivalent resistance of the parallel network is given by

$$\left(\frac{40 \times 160}{40 + 160}\right) \text{k}\Omega, \text{ i.e. } \left(\frac{40 \times 160}{200}\right) \text{k}\Omega = 32 \text{ k}\Omega .$$

The circuit is now effectively as shown in *Fig 12(b)*.
Thus the voltage indicated on the voltmeter is

$$\left(\frac{32}{32+60}\right) 100 \text{ V} = \mathbf{34.78\ V}$$

A considerable error is thus caused by the loading effect of the voltmeter on the circuit. The error is reduced by using a voltmeter with a higher sensitivity.

Problem 15 (a) A current of 20 A flows through a load having a resistance of 2 Ω. Determine the power dissipated in the load. (b) A wattmeter. whose current coil has a resistance of 0.01 Ω is connected as shown in *Fig 13*. Determine the wattmeter reading.

Fig 13

(a) Power dissipated in the load, $P = I^2R = (20)^2(2) = \mathbf{800\ W}$

(b) With the wattmeter connected in the circuit the total resistance R_T is 2+0.01 = 2.01 Ω

The wattmeter reading is thus $I^2R_T = (20)^2(2.01) = \mathbf{804\ W}$

Problem 16 The current flowing through a resistor of 5 kΩ ± 0.4% is measured as 2.5 mA with an accuracy of measurement of ± 0.5%. Determine the nominal value of the voltage across the resistor and its accuracy.

Voltage, $V = IR = (2.5 \times 10^{-3})(5 \times 10^3) = 12.5$ V
The maximum possible error is 0.4%+0.5% = 0.9%.
Hence the voltage, V = 12.5 V ± 0.9%. 0.9% of 12.5 V = 0.9/100 × 12.5
= 0.1125 V = 0.11 V correct to 2 significant figures. Hence the voltage V may also be expressed as **12.5 ± 0.11 volts** (i.e. a voltage lying between 12.39 V and 12.61 V).

Problem 17 The current I flowing in a resistor R is measured by a 0–10 A ammeter which gives an indication of 6.25 A. The voltage V across the resistor is measured by a 0–50 V voltmeter which gives an indication of 36.5 V. Determine the resistance of the resistor, and its accuracy of measurement if both instruments have a limit of error of 2% of FSD. Neglect any loading effects of the instruments.

Resistance, $R = \dfrac{V}{I} = \dfrac{36.5}{6.25} = 5.84\ \Omega$
Voltage error is ± 2% of 50 V = ± 1.0 V and expressed as a percentage of the voltmeter reading gives
$\dfrac{\pm 1}{36.5} \times 100\% = \pm 2.74\%$.

Current error is ± 2% of 10 A = ± 0.2 A and expressed as a percentage of the ammeter reading gives
$\dfrac{\pm 0.2}{6.25} \times 100\% = \pm 3.2\%$.

Maximum relative error = sum of errors = 2.74%+3.2% = ± 5.94%,
5.94% of 5.84 Ω = 0.347 Ω.
Hence the resistance of the resistor may be expressed as: **5.84 Ω ± 5.94%**,
or **5.84 ± 0.35 Ω** (rounding off).

Problem 18 The arms of a Wheatstone bridge ABCD have the following resistances: AB: R_1 = 1000 Ω ± 1.0%; BC: R_2 = 100 Ω ± 0.5%; CD: unknown resistance R_x; DA: R_3 = 432.5 Ω ± 0.2%. Determine the value of the unknown resistance and its accuracy of measurement.

The Wheatstone bridge network is shown in *Fig 14* and at balance: $R_1 R_x = R_2 R_3$

i.e. $R_x = \dfrac{R_2 R_3}{R_1} = \dfrac{(100)(432.5)}{1000} = 43.25\ \Omega$

The maximum relative error of R_x is given by the sum of the three individual errors, i.e.
1.0%+0.5%+0.2% = 1.7%
Hence R_x = 43.25 Ω ± 1.7%
1.7% of 43.25 Ω = 0.74 Ω (rounding off).
Thus R_x may also be expressed as
R_x = 43.25 ± 0.74 Ω

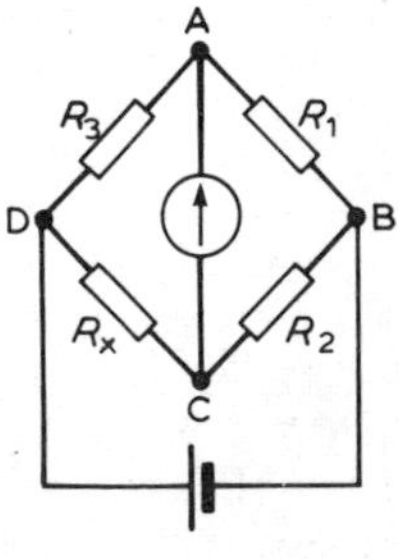

A
R_3
R_1
D
B
R_x
R_2
C

Fig 14

C. FURTHER PROBLEMS ON MEASURING INSTRUMENTS AND MEASUREMENTS

SHORT ANSWER PROBLEMS

1 Express the ratio of two powers P_1 and P_2 in decibel units.
2 What does a power level unit of dBm indicate?
3 Name five quantities that a CRO is capable of measuring.
4 Sketch a Wheatstone bridge circuit used for measuring an unknown resistance in a d.c. circuit and state the balance condition.
5 Name five types of a.c. bridge used for measuring unknown inductance, capacitance or resistance.
6 What is a universal bridge?
7 State the name of an a.c. bridge used for measuring inductance.
8 Briefly describe how the measurement of Q-factor may be achieved.
9 What is harmonic analysis?
10 What is a feature of waveforms containing the fundamental and odd harmonics?
11 Name two advantages of electronic measuring instruments compared with moving-coil or moving-iron instruments.
12 Why do instrument errors occur when measuring complex waveforms?

MULTI-CHOICE PROBLEMS (answers on page 191)

1 The input and output powers of a system are 2 mW and 18 mW respectively. The decibel power ratio of output power to input power is (a) 9; (b) 9.54; (c) 1/9; (d) 19.08
2 The input and output voltages of a system are 500 μV and 500 mV respectively. The decibel voltage ratio of output to input voltage (assuming input resistance equals load resistance) is (a) 1000; (b) 30; (c) 0; (d) 60.
The input and output currents of a system are 3 mA and 18 mA respectively. The decibel current ratio of output to input current (assuming the input and load resistances are equal) is:
(a) 15.56; (b) 6; (c) 1/6; (d) 7.78.
4 Which of the following statements is false?
(a) The Schering bridge is normally used for measuring unknown capacitances.
(b) A.c. electronic measuring instruments can handle a much wider range of frequency than the moving coil instrument.
(c) A complex waveform is one which is non-sinusoidal.
(d) A square-wave normally contains the fundamental and even harmonics.
5 A voltmeter has a FSD of 100 V, a sensitivity of 1 kΩ/V and an accuracy of ± 2% of FSD. When the voltmeter is connected into a circuit it indicates 50 V. Which of the following statements is false?
(a) Voltage reading is 50 ± 2 V; (b) Voltage resistance is 100 kΩ;
(c) Voltage reading is 50 V ± 2%; (d) Voltage reading is 50 V ± 4%.

Fig 15 shows double-beam CRO waveform traces. For the quantities stated in *Problems 6 to 12*, select the correct answer from the following list.

(a) 30 V; (b) 0.2 s; (c) 50 V; (d) $\frac{15}{\sqrt{2}}$ V; (e) 54° leading;

(f) $\frac{250}{\sqrt{2}}$ V; (g) 15 V; (h) 100 μs; (i) $\frac{50}{\sqrt{2}}$ V; (j) 250 V;

(k) 10 kHz; (l) 75 V; (m) 40 μs; (n) $\frac{3\pi}{10}$ rad lagging; (o) $\frac{25}{\sqrt{2}}$ V;

(p) 5 kHz; (q) $\frac{30}{\sqrt{2}}$ V; (r) 25 kHz; (s) $\frac{75}{\sqrt{2}}$ V; (t) $\frac{3\pi}{10}$ rad leading.

6 Amplitude of waveform P.
7 Peak-to-peak value of waveform Q.
8 Periodic time of both waveforms.
9 Frequency of both waveforms.
10 r.m.s. value of waveform P.
11 r.m.s. value of waveform Q.
12 Phase displacement of waveform Q relative to waveform P.

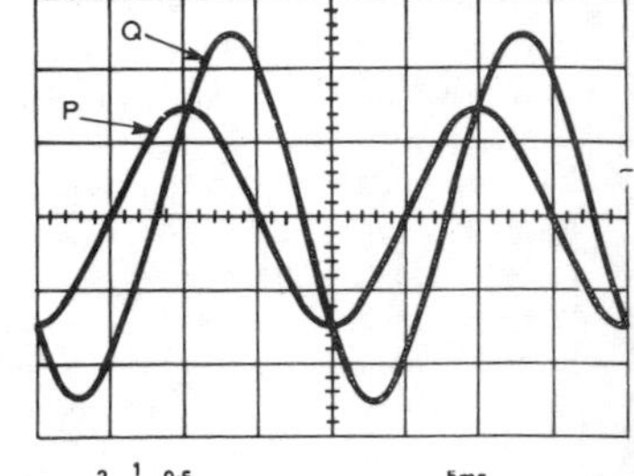

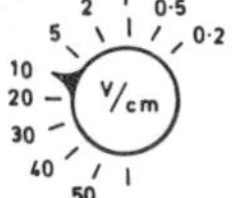

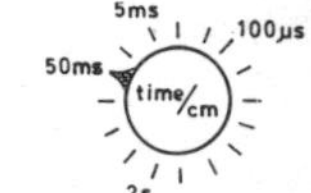

Fig 15

CONVENTIONAL PROBLEMS

1 The ratio of two powers is (a) 3; (b) 10; (c) 20; (d) 10 000. Determine the decibel power ratio for each. [(a) 4.77 dB; (b) 10 dB; (c) 13 dB; (d) 40 dB]

2 The ratio of two powers is (a) $\frac{1}{10}$; (b) $\frac{1}{3}$; (c) $\frac{1}{40}$; (d) $\frac{1}{100}$.

Determine the decibel power ratio for each.
[(a) –10 dB; (b) –4.77 dB; (c) –16.02 dB; (d) –20 dB]

3 The input and output currents of a system are 2 mA and 10 mA respectively. Determine the decibel current ratio of output to input current assuming input and output resistances of the system are equal. [13.98 dB]

4 5% of the power supplied to a cable appears at the output terminals. Determine the power loss in decibels. [13 dB]

5 An amplifier has a gain of 24 dB. Its input power is 10 mW. Find its output power. [2.51 W]

6 Determine, in decibels, the ratio of the output power to input power of a 4 stage system, the stages having gains of 10 dB, 8 dB, –5 dB and 7 dB. Find also the overall power gain. [20 dB; 100]

7 The output voltage from an amplifier is 7 mV. If the voltage gain is 25 dB calculate the value of the input voltage assuming that the amplifier input resistance and load resistance are equal. [0.39 mV]

8 The voltage gain of a number of cascaded amplifiers are 23 dB, –5.8 dB, –12.5 dB and 3.8 dB. Calculate the overall gain in decibels assuming that input and load resistances for each stage are equal. If a voltage of 15 mV is applied to the input of the system, determine the value of the output voltage.
[8.5 dB; 39.91 mV.]

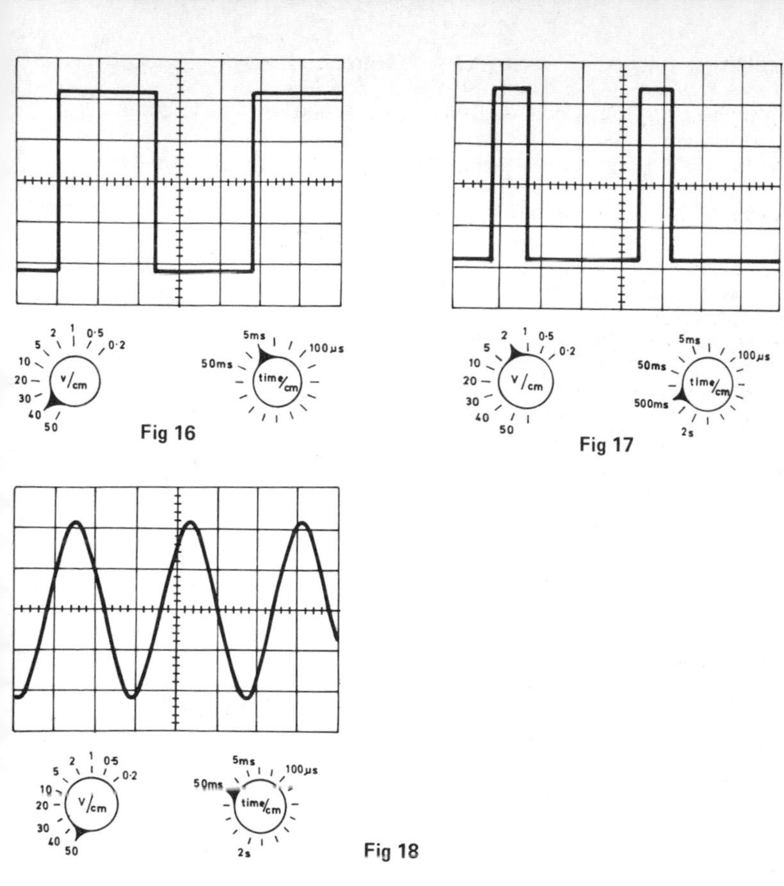

Fig 16

Fig 17

Fig 18

9 The scale of a voltmeter has a decibel scale added to it, which is calibrated by taking a reference level of 0 dB when a power of 1 mW is dissipated in a 600 Ω resistor. Determine the voltage at (a) 0 dB; (b) 1.5 dB and (c) –15 dB. (d) What decibel reading corresponds to 0.5 V?

[(a) 0.775 V; (b) 0.921 V; (c) 0.138 V; (d) –3.807 dB]

10 For the square voltage waveform displayed on a CRO shown in *Fig 16*, find (a) its frequency and (b) its peak to peak voltage.

[(a) 41.7 Hz; (b) 176 V]

11 For the pulse waveform shown in *Fig 17*, find (a) its frequency and (b) the magnitude of the pulse voltage. [(a) 0.56 Hz; (b) 8.4 V]

12 For the sinusoidal waveform shown in *Fig 18*, determine (a) its frequency; (b) the peak to peak voltage and (c) the r.m.s. voltage.

[(a) 7.14 Hz; (b) 220 V; (c) 77.78 V]

13 A Maxwell bridge circuit ABCD has the following arm impedances: AB, 250 Ω resistance; BC, 15 μF capacitor in parallel with a 10 kΩ resistor; CD, 400 Ω

resistor; DA, unknown inductor having inductance L and resistance R. Determine the values of L and R assuming the bridge is balanced. [1.5 H; 10 Ω]

14 A Q-meter measures the Q-factor of a series L–C–R circuit to be 200 at a resonant frequency of 250 kHz. If the capacitance of the Q-meter capacitor is set to 300 pF determine (a) the inductance L and (b) the resistance R of the inductor. [(a) 1.351 mH; (b) 10.61 Ω]

15 (a) A current of 15 A flows through a load having a resistance of 4 Ω. Determine the power dissipated in the load. (b) A wattmeter, whose current coil has a resistance of 0.02 Ω is connected (as shown in *Fig 13*) to measure the power in the load. Determine the wattmeter reading assuming the current in the load is still 15 A. [(a) 900 W; (b) 904.5 W]

16 A voltage of 240 V is applied to a circuit consisting of an 800 resistor in series with a 1.6 kΩ resistor. What is the voltage across the 1.6 kΩ resistor. The p.d. across the 1.6 kΩ resistor is measured by a voltmeter of FSD 250 V and sensitivity 100 Ω/V. Determine the voltage indicated. [160 V; 156.7 V]

17 The p.d. across a resistor is measured as 37.5 V with an accuracy of ± 0.5%. The value of the resistor is 6 kΩ ± 0.8%. Determine the current flowing in the resistor and its accuracy of measurement.
[6.25 mA ± 1.3% or 6.25 ± 0.08 mA]

18 The voltage across a resistor is measured by a 75 V FSD voltmeter which gives an indication of 52 V. The current flowing in the resistor is measured by a 20 A FSD ammeter which gives an indication of 12.5 A. Determine the resistance of the resistor and its accuracy if both instruments have an accuracy of ± 2% of FSD. [4.16 Ω ± 6.08% or 4.16 ± 0.25 Ω]

19 A Wheatstone bridge PQRS has the following arm resistances: PQ, 1 kΩ ± 2%; QR, 100 Ω ± 0.5%; RS, unknown resistance; SP, 273.6 Ω ± 0.1%. Determine the value of the unknown resistance, and its accuracy of measurement.
[27.36 Ω ± 2.6% or 27.36 ± 0.71 Ω]

10 Introduction to simple filter and attenuation circuits

A. MAIN POINTS CONCERNED WITH SIMPLE FILTER AND ATTENUATION CIRCUITS

1 A **filter** is a network designed to pass signals having frequencies within certain bands (called **passbands**) with little attenuation, but greatly attenuates signals within other bands (called **attenuation bands** or **stopbands**).
2 A filter is frequency sensitive and is thus composed of reactive elements. Since certain frequencies are to be passed with minimal loss, ideally the inductors and capacitors need to be pure components since the presence of resistance results in some attenuation at all frequencies.
3 Between the pass band of a filter, where ideally the attenuation is zero, and the attenuation band, where ideally the attenuation is infinite, is the **cut off frequency**, this being the frequency at which the attenuation changes from zero to some finite value.
4 A filter network containing no source of power is termed **passive** and one containing one or more power source is known as an **active** filter network.
5 Filters are used for a variety of purposes in nearly every type of electronic communications and control equipment. The bandwidths of filters used in communications systems vary from a fraction of a hertz to many megahertz, depending on the application. There are four basic types of filter sections, these being:
(a) low-pass
(b) high-pass
(c) band-pass
(d) band-stop
(See *problems 1 to 4*)
6 An **attenuator** is a device for introducing a specified loss between a signal source and a matched load without upsetting the impedance relationship necessary for matching. The loss introduced is constant irrespective of frequency; since reactive elements (L or C) vary with frequency, it follows that ideal attenuators are networks containing pure resistances. A fixed attenuator section is usually known as a 'pad'.
7 **Attenuation** is a reduction in the magnitude of a voltage or current due to its transmission over a line or through an attenuator. Any degree of attenuation may be achieved with an attenuator by suitable choice of resistance values but the input and output impedances of the pad must be such that the impedance conditions existing in the circuit into which it is connected are not disturbed.

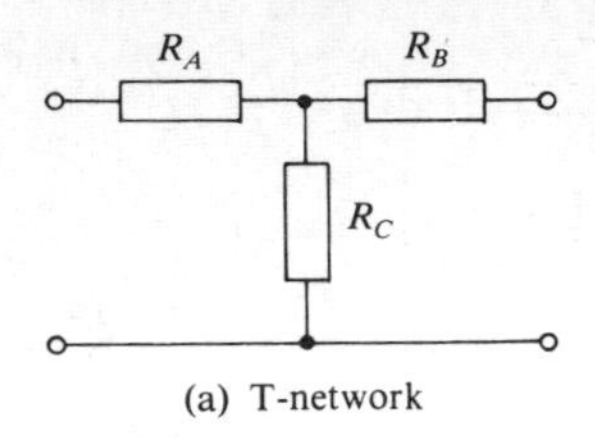

(a) T-network

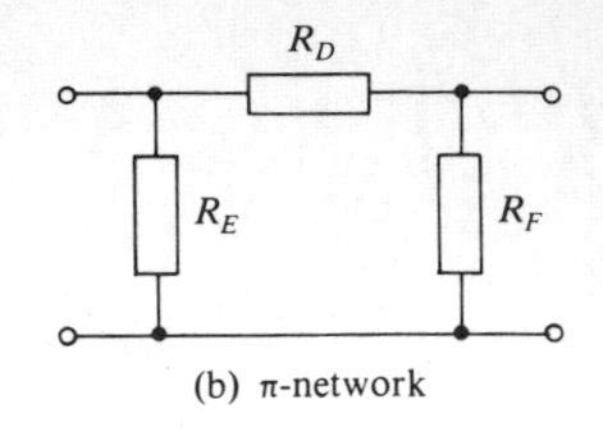

(b) π-network

Fig 1

Thus an attenuator must provide the correct input and output impedances as well as providing the required attenuation.

8 Attenuator sections are made up of resistances connected as T or π arrangements. *Fig 1(a)* shows a T-network, which is termed **symmetrical** if $R_A = R_B$ and *Fig 1(b)* shows a π-network which is symmetrical if $R_E = R_F$. If $R_A \neq R_B$ in *Fig 1(a)* and $R_E \neq R_F$ in *Fig 1(b)*, the sections are termed **asymmetrical**. Both networks shown have one common terminal, which may be earthed, and are therefore said to be **unbalanced**. The **balanced** form of the T-network is shown in *Fig 2(a)* and the balanced form of the π-network is shown in *Fig 2(b)*.

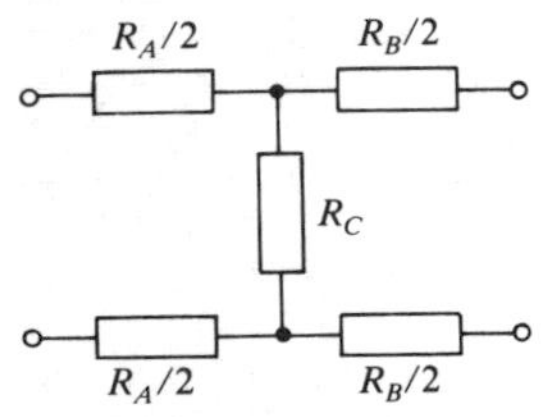

(a) Balanced T-network

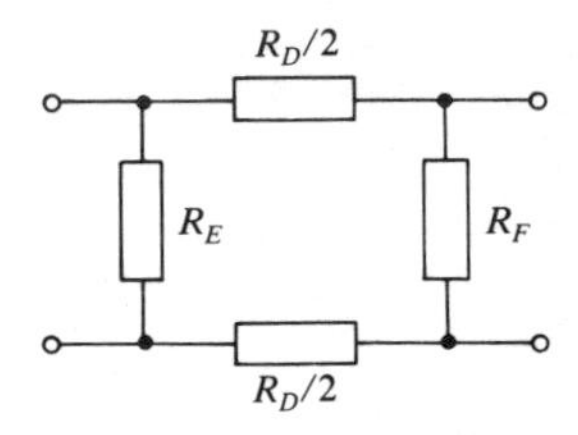

(b) Balanced π-network

Fig 2

9 Networks in which electrical energy is fed in at one pair of terminals and taken out at a second pair of terminals are called **two-port networks**.

For any passive two-port network it is found that a particular value of load impedance can always be determined which will produce an input impedance having the same value as the load impedance. This is called the **iterative impedance** for an asymmetrical network and its value depends on which pair of terminals is taken to be the input and which is the output; there are thus two values of iterative impedance, one for each direction. For a symmetrical network, the two iterative impedances are equal and this value is called the **characteristic impedance**.

10 For the **symmetrical T-attenuator** shown in *Fig 3*, the characteristic impedance R_o is given by:

$$R_o = \sqrt{(R_1^2 + 2R_1R_2)}$$

or $$R_o = \sqrt{(R_{oc}R_{sc})}$$

where R_{oc} = input resistance when the output is open–circuited

R_{sc} = input resistance when the output is short–circuited

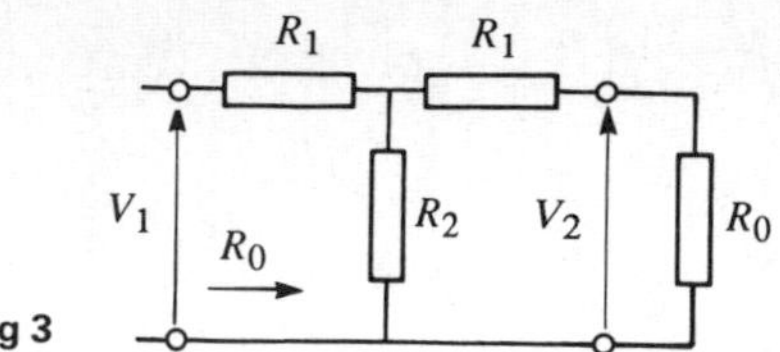

Fig 3

If the characteristic impedance R_o and the attenuation $N \left(=\frac{V_1}{V_2}\right)$ are known for the T-attenuator of *Fig 3*, then:

$$R_1 = R_o \left(\frac{N-1}{N+1}\right) \text{ and } R_2 = R_o \left(\frac{2N}{N^2-1}\right)$$

(See *problems 5 to 9*)

11 For the **symmetrical π-attenuator** shown in *Fig 4*, the characteristic impedance R_o is given by:

$$R_o = \sqrt{\left(\frac{R_1 R_2^2}{R_1 + 2R_2}\right)}$$

or $R_o = \sqrt{(R_{oc}R_{sc})}$

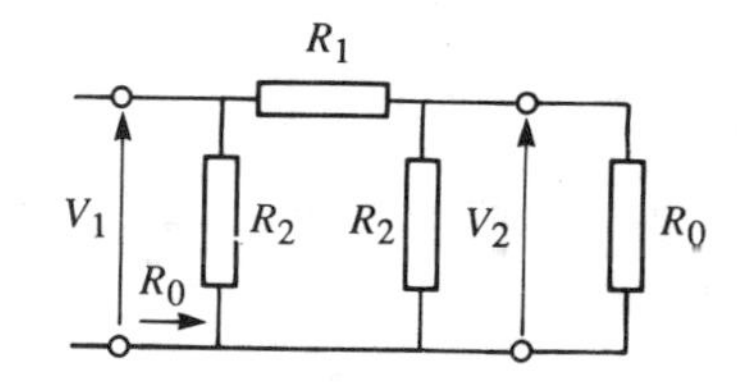

Fig 4

If the characteristic impedance R_o and the attenuation $N \left(=\frac{V_1}{V_2}\right)$ are known for the π-attenuator of *Fig 4*, then:

$$R_1 = R_o \left(\frac{N^2-1}{2N}\right) \text{ and } R_2 = R_o \left(\frac{N+1}{N-1}\right)$$

(See *problems 10 to 14*)

12 Often two-port networks are connected in **cascade**, i.e. the output from the first network becomes the input in the second network, and so on, as shown in *Fig 5*. Thus an attenuator may consist of several cascaded sections so as to achieve a particular desired overall performance.

If the cascade is arranged so that the impedance measured at one port and the impedance with which the other port is terminated have the same value, then each section (assuming they are symmetrical) will have the same characteristic impedance R_o and the last network will be terminated in R_o. Thus each network will have a matched termination and hence the attenuation in decibels of section 1 in *Fig 5* is given by $a_1 = 20 \lg (V_1/V_2)$. Similarly, the attenuation of

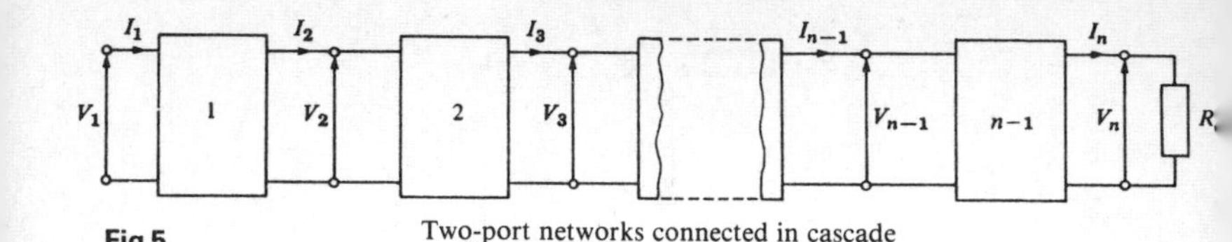

Fig 5 Two-port networks connected in cascade

section 2 is given by $a_2 = 20 \lg (V_2/V_3)$, and so on. The overall attenuation is given by:

$$a = 20 \lg \frac{V_1}{V_n}$$

$$= 20 \lg \left(\frac{V_1}{V_2} \times \frac{V_2}{V_3} \times \frac{V_3}{V_4} \times \ldots \times \frac{V_{n-1}}{V_n} \right)$$

$$= 20 \lg \frac{V_1}{V_2} + 20 \lg \frac{V_2}{V_3} + \ldots + 20 \lg \frac{V_{n-1}}{V_n}$$

by the laws of logarithms, i.e.,

overall attenuation, $a = a_1 + a_2 + \ldots + a_{n-1}$

Thus the overall attenuation is the sum of the attenuations (in decibels) of the matched sections.
(See *problems 15 and 16*)

B. WORKED PROBLEMS ON SIMPLE FILTER AND ATTENUATION CIRCUITS

Problem 1 Describe the function of a low-pass filter section. Sketch the ideal and practical attenuation/frequency characteristic. State examples where such a filter is used.

Fig 6 shows simple unbalanced T and π section filters using series inductors and shunt capacitors. If either section is connected into a network and a continuously increasing frequency is applied, each would have a frequency-attenuation characteristic as shown in *Fig 7(a)*. This is an ideal characteristic and assumes pure reactive elements. All frequencies are seen to be passed from zero up to a certain value without attenuation, this value being shown as f_c, the cut-off frequency; all values of frequency about f_c are attenuated. It is for this reason that the networks shown in *Figs 6(a) and (b)* are known as **low-pass filters**. The electrical circuit diagram symbol for a low-pass filter is shown in *Fig 7(b)*.

Summarizing, **a low-pass filter is one designed to pass signals at frequencies below a specified cut-off frequency.**

When rectifiers are used to produce the d.c. supplies of electronic systems, a large ripple introduces undesirable noise and may even mask the effect of the

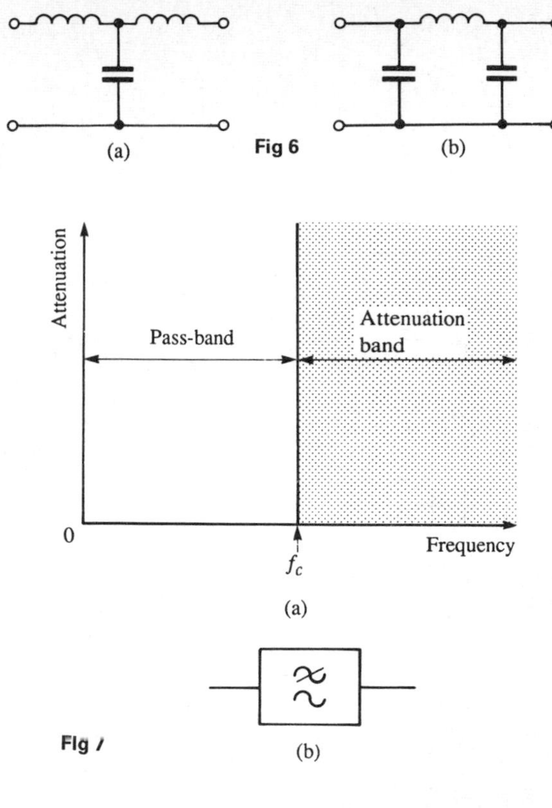

Fig 6

Fig 7

signal voltage. Low-pass filters are then added to smooth the output voltage waveform, this being one of the most common applications of filters in electrical circuits. Filters are employed to isolate various sections of a complete system and thus to prevent undesired interactions. For example, the insertion of low-pass decoupling filters between each of several amplifier stages and a common power supply reduces interaction due to the common power supply impedance.

Fig 7(a) shows an ideal low-pass filter section characteristic. In practice, the characteristic curve of a low-pass prototype filter section looks more like that shown in *Fig 8*. The characteristic may be improved somewhat closer to the ideal by connecting two or more identical sections in cascade. This produces a much sharper cut-off characteristic, although the attenuation in the pass band is increased a little.

Problem 2 Describe the function of a high-pass filter section and sketch the ideal and practical attenuation/frequency characteristic.

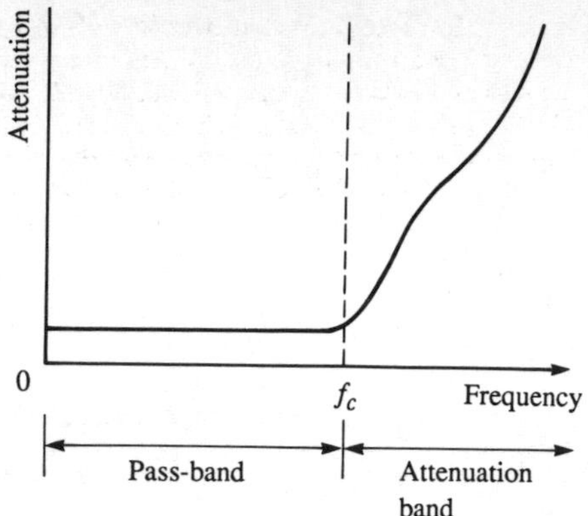

Fig 8

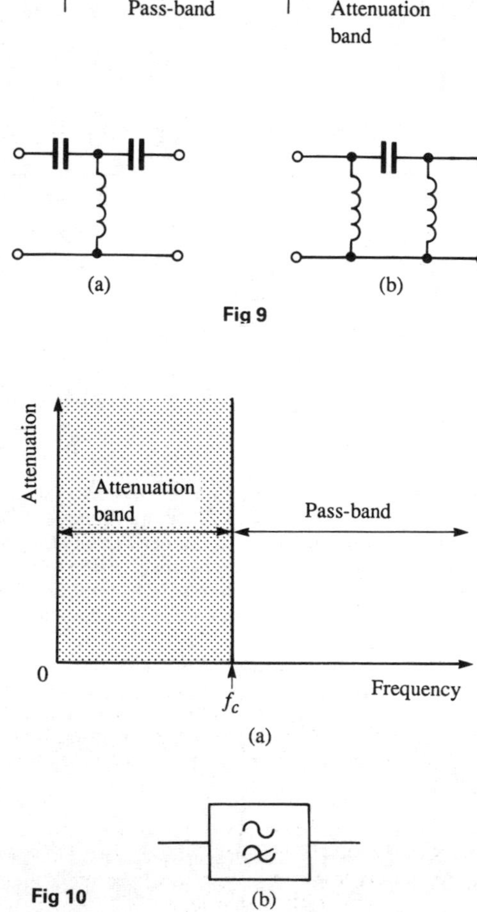

Fig 9

Fig 10

Fig 9 shows simple unbalanced T and π section filters using series capacitors and shunt inductors. If either section is connected into a network and a continuously increasing frequency is applied, each would have a frequency–attenuation characteristic as shown in *Fig 10(a)*. Once again this is an ideal characteristic assuming pure reactive elements. All frequencies below the cut-off frequency f_c are seen to be attenuated and all frequencies above f_c are passed without loss. It is for this reason that the networks shown in *Figs 9(a)* and *(b)* are known as **high-pass filters**. The electrical circuit-diagram symbol for a high-pass filter is shown in *Fig 10(b)*.

Summarizing, **a high-pass filter is one designed to pass signals at frequencies above a specified cut-off frequency**.

The characteristic shown in *Fig 10(a)* is ideal in that it is assumed that there is no attenuation at all in the pass-bands and infinite attenuation in the attenuation bands. Both of those conditions are impossible to achieve in practice. Due to resistance, mainly in the inductive elements the attenuation in the pass-band will not be zero, and in a practical filter section the attenuation in the attenuation band will have a finite value. In addition to the resistive loss there is often an added loss due to mismatching.

Fig 10(a) shows an ideal high-pass filter section characteristic of attenuation against frequency. In practice, the characteristic curve of a high-pass prototype filter section would look more like that shown in *Fig 11*.

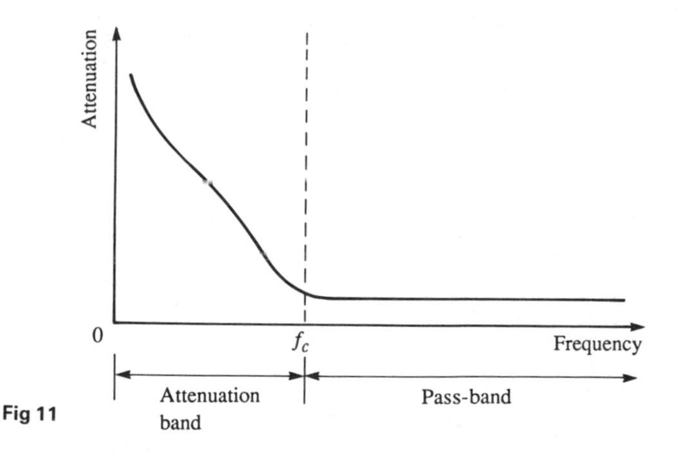

Fig 11

Problem 3 Describe the function of a band-pass filter section. State an application of such a filter and sketch ideal and practical attenuation/frequency characteristics.

A band-pass filter is one designed to pass signals with frequencies between two specified cut-off frequencies. The characteristic of an ideal band-pass filter is shown in *Fig 12*. Such a filter may be formed by cascading a high-pass and a low-pass filter. f_{CH} is the cut-off frequency of the high-pass filter and f_{CL} is the

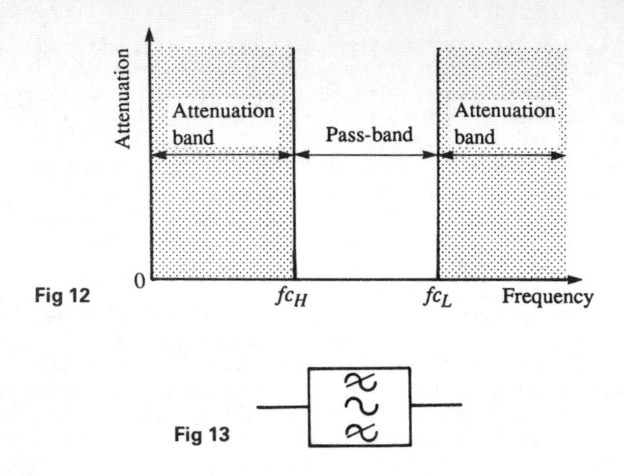

Fig 12

Fig 13

cut-off frequency of the low-pass filter. As can be seen, $f_{CL} > f_{CH}$ for a band-pass filter, the pass-band being given by the difference between these values. The electrical circuit diagram symbol for a band-pass filter is shown in *Fig 13*.

Crystal and ceramic devices are used extensively in band-pass filters. They are common in the intermediate-frequency amplifiers of vhf radios where a precisely defined bandwidth must be maintained for good performance.

A typical practical characteristic for a band-pass filter is shown in *Fig 14*.

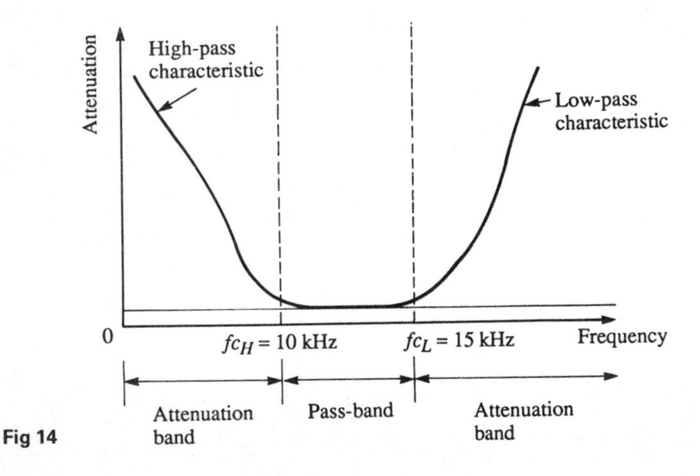

Fig 14

Problem 4 Describe the function of a band-stop filter section. State an application of such a filter and sketch the ideal and practical attenuation/frequency characteristic.

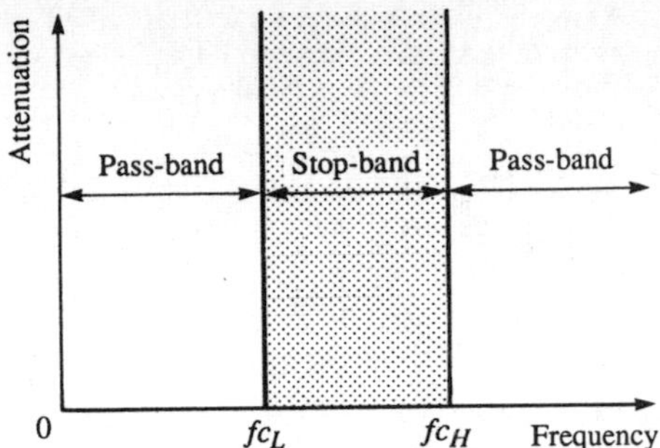

Fig 15

A band-stop filter is one designed to pass signals with all frequencies except those between two specified cut-off frequencies. The characteristic of an ideal band-stop filter is shown in *Fig 15*. Such a filter may be formed by connecting a high-pass and a low-pass filter in parallel. As can be seen, for a band-stop filter $f_{CH} > f_{CL}$, the band-stop being given by the difference between these values. The electrical circuit diagram symbol for a band-stop filter is shown in *Fig 16*.

Fig 16

Sometimes, as in the case of interference from 50 Hz power lines in an audio system, the exact frequency of a spurious noise signal is known. Usually such interference is from an odd harmonic of 50 Hz, for example, 250 Hz. A sharply tuned band-stop filter, designed to attenuate the 250 Hz noise signal, is used to minimize the effect of the output. A high-pass filter with cut-off frequency

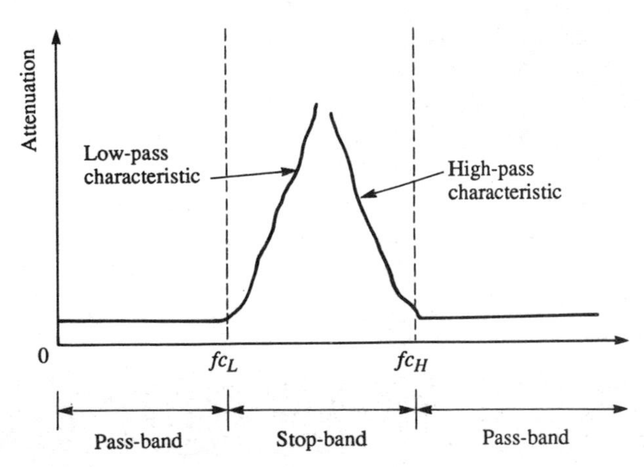

Fig 17

greater than 250 Hz would also remove the interference, but some of the lower frequency components of the audio signal would be lost as well.

A typical practical characteristic for a band-stop filter is shown in *Fig 17*.

Problem 5 A symmetrical T-network has series resistances R_1 and shunt resistance R_2. Show that the characteristic impedance R_o is given by (i) $R_o = \sqrt{(R_1^2 + 2R_1R_2)}$
and (ii) $R_o = \sqrt{(R_{oc}R_{sc})}$

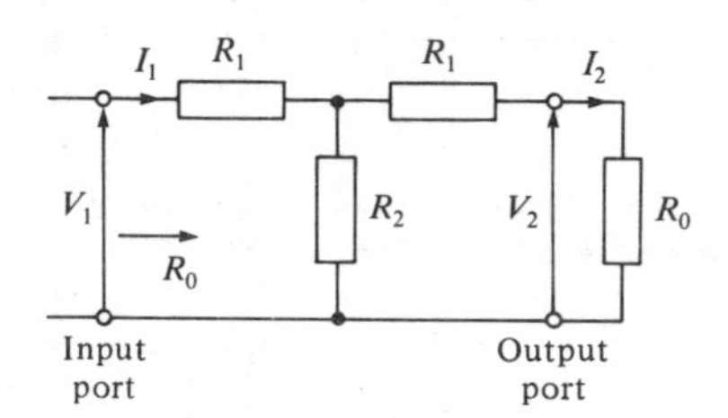

Fig 18

Fig 18 shows a symmetrical T-network terminated in an impedance R_o. If the resistance 'looking-in' at the input port is also R_o, then R_o is the characteristic impedance.

(i) From *Fig 18*,

$$\frac{V_1}{I_1} = R_o = \frac{V_2}{I_2}$$

Hence $\dfrac{V_1}{I_1} = R_o = R_1 + \dfrac{R_2(R_1 + R_o)}{R_2 + R_1 + R_o}$, since $(R_1 + R_o)$ is in parallel with R_2,

$$= \frac{R_1^2 + R_1R_2 + R_1R_o + R_1R_2 + R_2R_o}{R_1 + R_2 + R_o}$$

i.e. $R_o = \dfrac{R_1^2 + 2R_1R_2 + R_1R_o + R_2R_o}{R_1 + R_2 + R_o}$

Thus $R_o(R_1 + R_2 + R_o) = R_1^2 + 2R_1R_2 + R_1R_o + R_2R_o$

$$R_oR_1 + R_oR_2 + R_o^2 = R_1^2 + 2R_1R_2 + R_1R_o + R_2R_o$$

i.e. $R_o^2 = R_1^2 + 2R_1R_2$

from which, characteristic impedance, $\boldsymbol{R_o = \sqrt{(R_1^2 + 2R_1R_2)}}$ (1)

(ii) If the output terminals of *Fig 18* are open-circuited, then the open-circuit resistance, $R_{oc} = R_1 + R_2$.
If the output terminals of *Fig 18* are short-circuited, then the short-circuit resistance, $R_{sc} = R_1 + \dfrac{R_1R_2}{R_1 + R_2} = \dfrac{R_1^2 + 2R_1R_2}{R_1 + R_2}$

Thus $R_{oc}R_{sc} = (R_1 + R_2)\left(\dfrac{R_1^2 + 2R_1R_2}{R_1 + R_2}\right) = R_1^2 + 2R_1R_2$

Comparing with equation (1) gives: $\boldsymbol{R_o = \sqrt{(R_{oc}R_{sc})}}$ (2)

Problem 6 Determine the characteristic impedance of each of the attenuator sections shown in *Fig 19*.

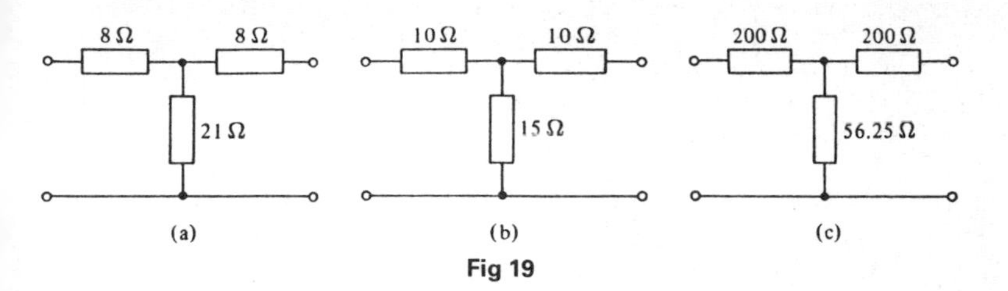

Fig 19

From para 10 and equation (1), *problem 5*, for a T-section attenuator the characteristic impedance, $R_o = \sqrt{(R_1^2 + 2R_1R_2)}$.

(a) $R_o = \sqrt{(8^2 + (2)(8)(21))} = \sqrt{400} = \mathbf{20\ \Omega}$

(b) $R_o = \sqrt{(10^2 + (2)(10)(15))} = \sqrt{400} = \mathbf{20\ \Omega}$

(c) $R_o = \sqrt{(200^2 + (2)(200)(56.25))} = \sqrt{62500} = \mathbf{250\ \Omega}$

It is seen that the characteristic impedance of parts (a) and (b) is the same. In fact, there are numerous combinations of resistances R_1 and R_2 which would give the same value for the characteristic impedance.

Problem 7 For the attenuator network shown in *Fig 20*, determine (a) the input resistance when the output port is open-circuited, (b) the input resistance when the output port is short-circuited, and (c) the characteristic impedance.

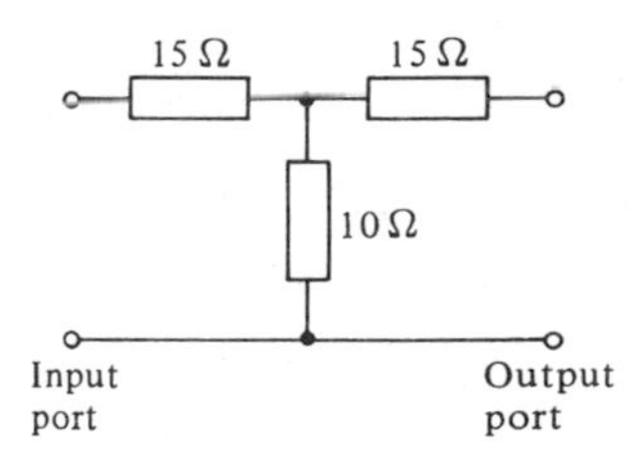

Fig 20

For the T-network shown in *Fig 20*:

(a) $R_{oc} = 15 + 10 = \mathbf{25\ \Omega}$

(b) $R_{sc} = 15 + \dfrac{10 \times 15}{10 + 15} = 15 + 6 = \mathbf{21\ \Omega}$

(c) From para 10 and equation 2, *problem 5*,

$$R_o = \sqrt{(R_{oc}R_{sc})} = \sqrt{[(25)(21)]} = \mathbf{22.9\ \Omega}$$

(Alternatively, $R_o = \sqrt{(R_1^2 + 2R_1R_2)} = \sqrt{(15^2 + (2)(15)(10))}$
$= \mathbf{22.9\ \Omega}$)

Problem 8 For a symmetrical T-network attenuator the characteristic impedance is R_o and the attenuation N. Determine expressions for series resistance R_1 and shunt resistance R_2 in terms of R_o and N.

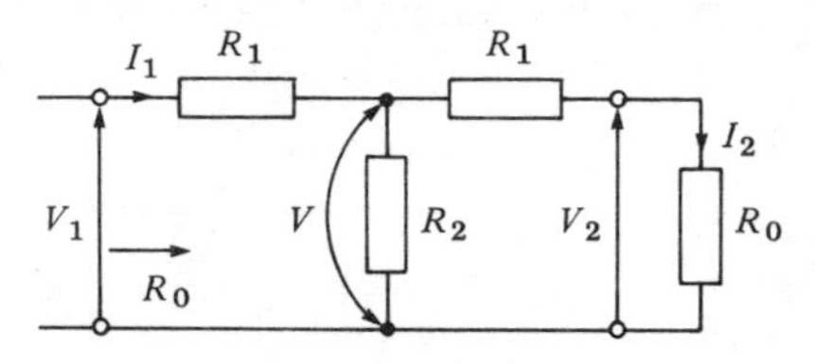

Fig 21 Symmetrical T-pad attenuator

A symmetrical T-network attenuator terminated in its characteristic impedance R_o is shown in *Fig 21*. Attenuation $N = \frac{V_1}{V_2}$

From *Fig 21*,

$$\text{current } I_1 = \frac{V_1}{R_o}$$

$$\text{Voltage } V = V_1 - I_1R_1 = V_1 - \left(\frac{V_1}{R_o}\right) R_I$$

i.e.,
$$V = V_1\left(1 - \frac{R_1}{R_o}\right)$$

$$\text{Voltage } V_2 = \left(\frac{R_o}{R_1 + R_o}\right) V, \text{ by voltage division}$$

i.e.,
$$V_2 = \left(\frac{R_o}{R_1 + R_o}\right) V_1 \left(1 - \frac{R_1}{R_o}\right) = V_1\left(\frac{R_o}{R_1 + R_o}\right)\left(\frac{R_o - R_1}{R_o}\right)$$

Hence
$$\frac{V_2}{V_1} = \frac{R_o - R_1}{R_o + R_1} \text{ or } \frac{V_1}{V_2} = N = \frac{R_o + R_1}{R_o - R_1}$$

$$N(R_o - R_1) = R_o + R_1$$
$$NR_o - NR_1 = R_o + R_1$$
$$R_o(N - 1) = R_1(1 + N)$$

from which
$$R_1 = R_o\frac{(N-1)}{(N+1)} \quad (3)$$

From para 10 and equation (1), *problem 5*,

$$R_o = \sqrt{(R_1^2 + 2R_1R_2)} \text{ i.e., } R_o^2 = R_1^2 + 2R_IR_2$$

from which $R_2 = \frac{R_o^2 - R_1^2}{2R_1}$

Substituting for R_1 from equation (3) gives

$$R_2 = \frac{R_o^2 - [R_o(N-1)/(N+1)]^2}{2[R_o(N-1)/(N+1)]} = \frac{[R_o^2(N+1)^2 - R_o^2(N-1)^2]/(N+1)^2}{2R_o(N-1)/(N+1)}$$

i.e.,

$$R_2 = \frac{R_o^2[(N+1)^2 - (N-1)^2]}{2R_o(N-1)(N+1)} = \frac{R_o[(N^2+2N+1) - (N^2-2N+1)]}{2(N^2-1)}$$

$$= \frac{R_o(4N)}{2(N^2-1)}$$

Hence $\quad \mathbf{R_2 = R_o \left(\frac{2N}{N_2 - 1}\right)}$ (4)

Problem 9 Design a T-section symmetrical attenuator pad to provide a voltage attenuation of 20 dB and having a characteristic impedance of 600 Ω.

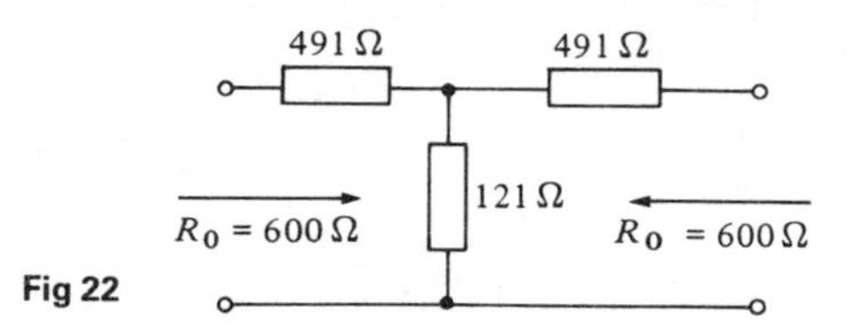

Fig 22

Voltage attenuation in decibels = 20 lg (V_1/V_2).
Attenuation, $N = V_1/V_2$, hence $20 = 20 \lg N$, from which $N = 10$.
Characteristic impedance, $R_o = 600\ \Omega$.
From equation (3), *problem 8*,

resistance $R_1 = \dfrac{R_o(N-1)}{(N+1)} = \dfrac{600(10-1)}{(10+1)} = \mathbf{491\ \Omega}$

From equation (4), *problem 8*,

resistance $R_2 = R_o\left(\dfrac{2N}{N^2-1}\right) = 600\left(\dfrac{(2)(10)}{10^2-1}\right) = \mathbf{121\ \Omega}$.

Thus the T-section attenuator shown in *Fig 22* has a voltage attenuation of 20 dB and a characteristic impedance of 600 Ω.
(Check: From para 10 and equation (1), *problem 5*,

$R_o = \sqrt{(R_1^2 + 2R_1R_2)} = \sqrt{[491^2 + 2(491)(121)]} = 600\ \Omega$)

Problem 10 A symmetrical π-network has series resistance R_1 and shunt resistance R_2. Show that the characteristic impedance R_o is given by

(i) $R_o = \left(\sqrt{\dfrac{R_1R_2^2}{R_1 + 2R_2}}\right)$

(ii) $R_o = \sqrt{(R_{oc}R_{sc})}$

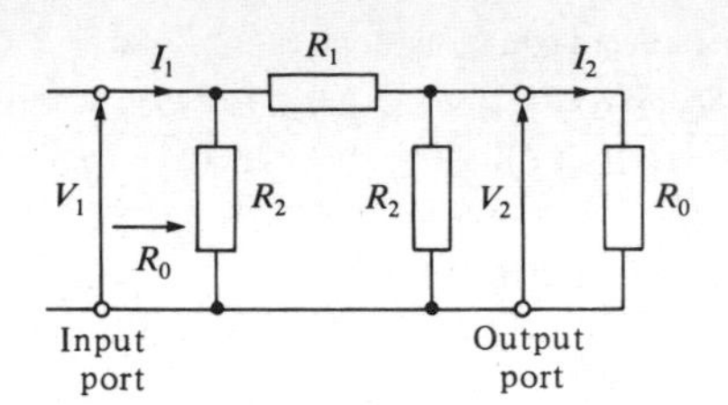

Fig 23

Figure 23 shows a symmetrical π-network terminated in an impedance R_o. If the resistance 'looking-in' at the input port is also R_o, then R_o is the characteristic impedance.

(i) From *Fig 23*,

$$\frac{V_1}{I_1} = R_o = (R_2) \text{ in parallel with } [R_1 \text{ in series with } (R_o \text{ and } R_2) \text{ in parallel}]$$

$$= (R_2) \text{ in parallel with } \left[R_1 + \frac{R_oR_2}{R_o + R_2}\right]$$

$$= (R_2) \text{ in parallel with } \left[\frac{R_1R_o + R_1R_2 + R_oR_2}{R_o + R_2}\right]$$

$$\text{i.e., } R_o = \frac{(R_2)\left[\frac{R_1R_o + R_1R_2 + R_oR_2}{R_o + R_2}\right]}{R_2 + \left[\frac{R_1R_o + R_1R_2 + R_oR_2}{R_o + R_2}\right]}$$

$$= \frac{\left(\frac{R_1R_2R_o + R_1R_2^2 + R_oR_2^2}{R_o + R_2}\right)}{\left(\frac{R_2R_o + R_2^2 + R_1R_o + R_1R_2 + R_oR_2}{R_o + R_2}\right)}$$

$$= \frac{R_1R_2R_o + R_1R_2^2 + R_oR_2^2}{R_2^2 + 2R_2R_o + R_1R_o + R_1R_2}$$

Thus $(R_o)(R_2^2 + 2R_2R_o + R_1R_o + R_1R_2) = R_1R_2R_o + R_1R_2^2 + R_oR_2^2$

i.e., $2R_2R_o^2 + R_1R_o^2 = R_1R_2^2$

and $R_o^2(2R_2 + R_1) = R_1R_2^2$

from which, characteristic impedance,

$$R_o = \sqrt{\left(\frac{R_1R_2^2}{R_1 + 2R_2}\right)} \quad (5)$$

(ii) If the output terminals of *Fig 23* are open-circuited, then the open-circuit resistance,

$$R_{oc} = \frac{R_2(R_1 + R_2)}{R_2 + R_1 + R_2} = \frac{R_2(R_1 + R_2)}{R_1 + 2R_2}$$

If the output terminals of *Fig 23* are short-circuited, then the short-circuit resistance,

$$R_{sc} = \frac{R_2 R_1}{R_1 + R_2}$$

Thus $$R_{oc} R_{sc} = \frac{R_2(R_1 + R_2)}{(R_1 + 2R_2)} \left(\frac{R_2 R_1}{R_1 + R_2} \right) = \frac{R_1 R_2^2}{R_1 + 2R_2}$$

Comparing with equation (5) gives:

$$R_o = \sqrt{(R_{oc} R_{sc})} \qquad (6)$$

From above, and from *problem 5*, it is seen that the characteristic impedance R_o is given by $R_o = \sqrt{(R_{oc} R_{sc})}$ whether the network is a symmetrical T or a symmetrical π.

Problem 11 Determine the characteristic impedance for the π section attenuator shown in *Fig 24*.

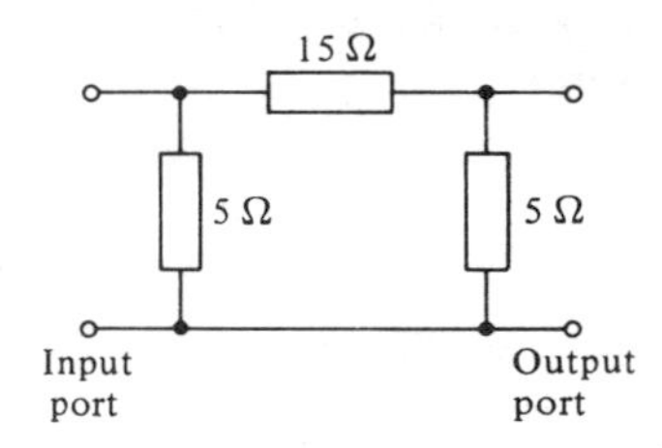

Fig 24

From para 11 and equation (5), *problem 10*,

characteristic impedance $R_o = \sqrt{\left(\frac{R_1 R_2^2}{R_1 + 2R_2} \right)} = \sqrt{\left(\frac{15(5)^2}{15 + 2(5)} \right)}$

i.e. $\mathbf{R_o = 3.87\ \Omega}$

Alternatively, $R_{oc} = \frac{5(15 + 5)}{5 + (15 + 5)} = \frac{100}{25} = 4\ \Omega$

$$R_{sc} = \frac{(5)(15)}{5 + 15} = \frac{75}{20} = 3.75\ \Omega$$

and $R_o = \sqrt{(R_{oc} R_{sc})} = \sqrt{[(4)(3.75)]} = \mathbf{3.87\ \Omega}$, as above.

Problem 12 A symmetrical π-attenuator pad has a series arm of 500 Ω resistance and each shunt arm of 1 kΩ resistance. Determine (a) the characteristic impedance, and (b) the attenuation (dB) produced by the pad.

The π-attenuator section is shown in *Fig 25* terminated in its characteristic impedance, R_o.

(a) From para 11 and equation (5), *problem 10*, for a symmetrical π-attenuator section,

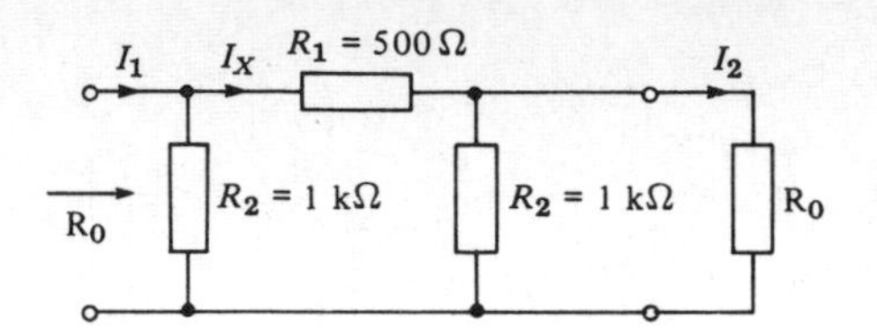

Fig 25

characteristic impedance, $R_o = \sqrt{\left(\dfrac{R_1R_2^2}{R_1 + 2R_2}\right)}$

Hence $R_o = \sqrt{\left(\dfrac{(500)(1000)^2}{500 + 2(1000)}\right)} = \mathbf{447\ \Omega}$

(b) From paras 1 and 2 of chapter 9, the power ratio in decibels

$$= 10 \lg \frac{P_2}{P_1}$$

and if P_1 and P_2 refer to power developed in two equal resistors then the power ratio in decibels

$$= 20 \lg \frac{V_2}{V_1} = 20 \lg \frac{I_2}{I_1}$$

Alternatively, attenuation $= 20 \lg \dfrac{V_1}{V_2} = 20 \lg \dfrac{I_1}{I_2}$

Thus from *Fig 25*,

attenuation $= 20 \lg \dfrac{I_1}{I_2}$

current $I_x = \left(\dfrac{R_2}{R_2 + R_1 + (R_2R_o/(R_2 + R_o))}\right)(I_1)$, by current division, i.e.

$$I_x = \left(\frac{1000}{1000 + 500 + ((1000)(447)/(1000 + 447))}\right) I_1$$

$$= 0.553\ I_1$$

and

current $I_2 = \left(\dfrac{R_2}{R_2 + R_o}\right) I_x = \left(\dfrac{1000}{1000 + 447}\right) I_x = 0.691\ I_x$

Hence $I_2 = 0.691(0.553\ I_1) = 0.382\ I_1$ and $I_l/I_2 = 1/0.382 = 2.617$
Thus **attenuation** $= 20 \lg 2.617 = \mathbf{8.36\ dB}$
(Alternatively, since $I_1/I_2 = N$, then the formula

$R_2 = R_o\left(\dfrac{N+1}{N-1}\right)$ (see para 11 and equation (7), *problem 13*) may be transposed for N, from which attenuation $= 20 \lg N$)

Problem 13 For a symmetrical π-network attenuator the characteristic impedance is R_o and the attenuation N. Determine expressions for series resistance R_1 and shunt resistance R_2 in terms of R_o and N.

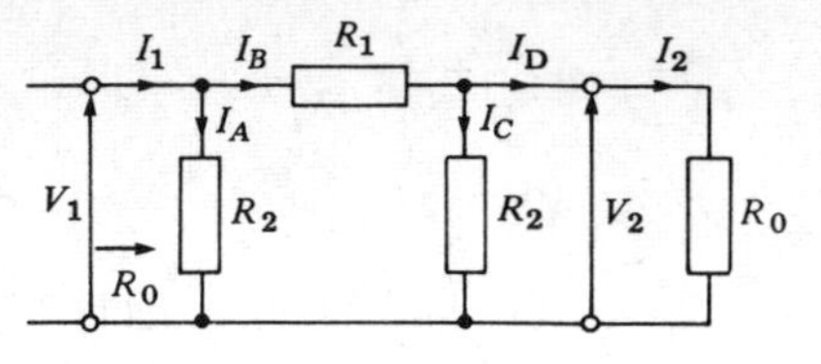

Fig 26 Symmetrical π-attenuator

A symmetrical π-network attenuator terminated in its characteristic impedance R_o is shown in *Fig 26*.

Attenuation $N = \dfrac{V_1}{V_2} \left(= \dfrac{I_1}{I_2}\right)$ from which $V_2 = \dfrac{V_1}{N}$

From *Fig 26*, current $I_1 = I_A + I_B$ and current $I_B = I_C + I_D$. Thus

current
$$I_1 = \frac{V_1}{R_o} = I_A + I_C + I_D$$

$$= \frac{V_1}{R_2} + \frac{V_2}{R_2} + \frac{V_2}{R_o} = \frac{V_1}{R_2} + \frac{V_1}{NR_2} + \frac{V_1}{NR_o}$$

since $V_2 = V_1/N$, i.e.,

$$\frac{V_1}{R_o} = V_1\left(\frac{1}{R_2} + \frac{1}{NR_2} + \frac{1}{NR_o}\right)$$

Hence

$$\frac{1}{R_o} = \frac{1}{R_2} + \frac{1}{NR_2} + \frac{1}{NR_o}$$

$$\frac{1}{R_o} - \frac{1}{NR_o} = \frac{1}{R_2} + \frac{1}{NR_2}$$

$$\frac{1}{R_o}\left(\frac{N-1}{N}\right) = \frac{1}{R_2}\left(\frac{N+1}{N}\right)$$

$$\frac{1}{R_o}\left(\frac{N-1}{N}\right) = \frac{1}{R_2}\left(\frac{N+1}{N}\right)$$

Thus

$$\mathbf{R_2 = R_o \frac{(N+1)}{(N-1)}} \qquad (7)$$

From *Fig 26*, current $I_1 = I_A + I_B$, and since the p.d. across R_1 is $(V_1 - V_2)$,

$$\frac{V_1}{R_o} = \frac{V_1}{R_2} + \frac{V_1 - V_2}{R_1}$$

$$\frac{V_1}{R_o} = \frac{V_1}{R_2} + \frac{V_1}{R_1} - \frac{V_2}{R_1}$$

$$\frac{V_1}{R_o} = \frac{V_1}{R_2} + \frac{V_1}{R_1} - \frac{V_1}{NR_1} \text{ since } V_2 = V_1/N$$

$$\frac{1}{R_o} = \frac{1}{R_2} + \frac{1}{R_1} - \frac{1}{NR_1}$$

$$\frac{1}{R_o} - \frac{1}{R_2} = \frac{1}{R_1}\left(1 - \frac{1}{N}\right)$$

$$\frac{1}{R_o} - \frac{(N-1)}{R_o(N+1)} = \frac{1}{R_1}\left(\frac{N-1}{N}\right) \text{ from equation (7),}$$

$$\frac{1}{R_o}\left(1 - \frac{N-1}{N+1}\right) = \frac{1}{R_1}\left(\frac{N-1}{N}\right)$$

$$\frac{1}{R_o}\left(\frac{(N+1)-(N-1)}{(N+1)}\right) = \frac{1}{R_1}\left(\frac{N-1}{N}\right)$$

$$\frac{1}{R_o}\left(\frac{2}{N+1}\right) = \frac{1}{R_1}\left(\frac{N-1}{N}\right)$$

$$R_1 = R_o\left(\frac{N-1}{N}\right)\left(\frac{N+1}{2}\right)$$

Hence

$$\mathbf{R_1 = R_o\left(\frac{N^2-1}{2N}\right)} \qquad (8)$$

Problem 14 Design a π-section symmetrical attenuator pad to provide a voltage attenuation of 20 dB and having a characteristic impedance of 600 Ω.

From problem 9, $N = 10$ and $R_o = 600\ \Omega$
From equation (8), *problem 13*,

$$\text{resistance } R_1 = R_o\left(\frac{N^2-1}{2N}\right) = 600\left(\frac{10^2-1}{(2)(10)}\right) = \mathbf{2970\ \Omega \text{ or } 2.97\ k\Omega}.$$

From equation (7), *problem 13*,

$$R_2 = R_o\left(\frac{N+1}{N-1}\right) = 600\left(\frac{10+1}{10-1}\right) = \mathbf{733\ \Omega}.$$

Thus the π-section attenuator shown in *Fig 27* has a voltage attenuation of 20 dB and a characteristic impedance of 600 Ω.
(Check: From equation (15)),

$$R_o = \sqrt{\left(\frac{R_1R_2^2}{R_1+2R_2}\right)} = \sqrt{\left(\frac{(2970)(733)^2}{2970+(2)(733)}\right)} = 600\ \Omega$$

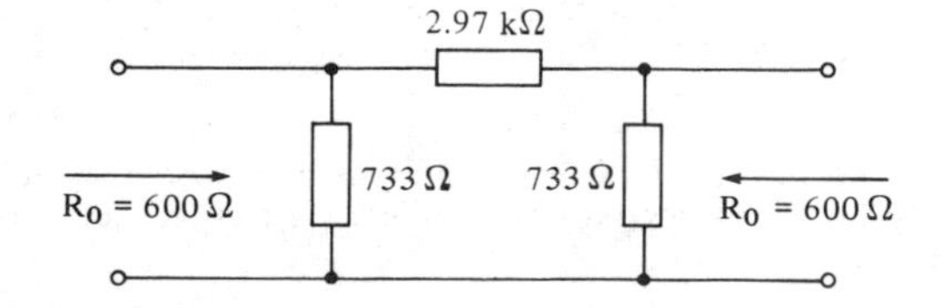

Fig 27

Problem 15 Five identical attenuator sections are connected in cascade. The overall attenuation is 70 dB and the voltage input to the first section is 20 mV. Determine
(a) the attenuation of each individual attenuator section,
(b) the voltage output of the final stage, and
(c) the voltage output of the third stage.

(a) From para 12, the overall attenuation is equal to the sum of the attenuations of the individual sections and, since in this case each section is identical, **the attenuation of each section** = 70/5 = **14 dB**

(b) If V_1 = the input voltage to the first stage and V_o = the output voltage of the final stage, then the overall attenuation = $20 \lg (V_1/V_o)$, i.e.,

$$70 = 20 \lg \left(\frac{20}{V_o}\right) \quad \text{where } V_o \text{ is in millivolts}$$

$$3.5 = \lg \left(\frac{20}{V_o}\right)$$

$$10^{3.5} = \frac{20}{V_o}$$

from which
output voltage of final stage,

$$V_o = \frac{20}{10^{3.5}} = 6.32 \times 10^{-3} \text{ mV} = \mathbf{6.32\ \mu V}$$

(c) The overall attenuation of three identical stages is 3 × 14 = 42 dB. Hence $42 = 20 \lg (V_1/V_3)$, where V_3 is the voltage output of the third stage. Thus

$$\frac{42}{20} = \lg \left(\frac{20}{V_3}\right) \text{ and } 10^{42/20} = \frac{20}{V_3}$$

from which **the voltage output of the third stage,**

$$V_3 = 20/10^{2.1} = \mathbf{0.159\ mV}$$

Problem 16 A d.c. generator has an internal resistance of 450 Ω and supplies a 450 Ω load.
(a) Design a T-network attenuator pad having a characteristic impedance of 450 Ω which, when connected between the generator and the load, will reduce the load current to $\frac{1}{8}$ of its initial value.
(b) If two such networks as designed in (a) were connected in series between the generator and the load, determine the fraction of the initial current that would now flow in the load.
(c) Determine the attenuation in decibels given by four such sections as designed in (a).

The T-network attenuator is shown in *Fig 28* connected between the generator and the load. Since it is matching equal impedances, the network is symmetrical.

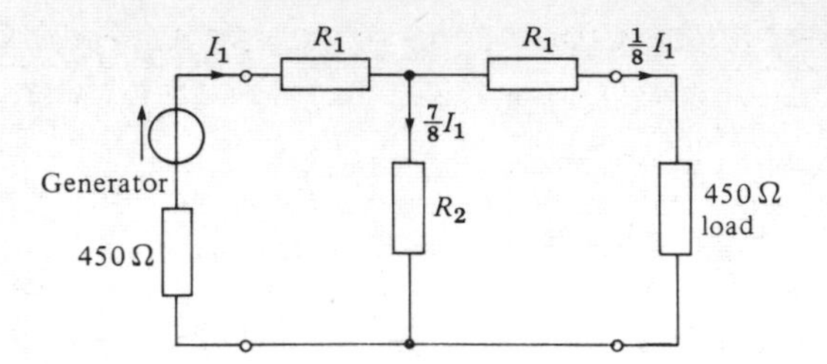

Fig 28

(a) Since the load current is to be reduced to $\frac{1}{8}$ of its initial value, the attenuation, $N = 8$. From equation (3),

$$\text{resistance, } R_1 = \frac{R_o(N-1)}{(N+1)} = 450\,\frac{(8-1)}{(8+1)} = \mathbf{350\ \Omega}$$

and from equation (4),

$$\text{resistance, } R_2 = R_o\left(\frac{2N}{N^2-1}\right) = 450\left(\frac{2\times 8}{8^2-1}\right) = \mathbf{114\ \Omega}$$

(b) When two such networks are connected in series, as shown in *Fig 29*, current I_1 flows into the first stage and $\frac{1}{8}\,I_1$ flows out of the first stage into the second. Again, $\frac{1}{8}$ of this current flows out of the second stage, i.e., $\frac{1}{8}\times\frac{1}{8}\,I_1$, i.e., 1/64 of I_1 flows into the load. **Thus 1/64 of the original current flows in the load.**

(c) The attenuation of a single stage is 8. Expressed in decibels, the attenuation is $20\lg(I_1/I_2) = 20\lg 8 = 18.06$ dB. From para 12, the overall attenuation of four identical stages is given by 18.06 + 18.06 + 18.06 + 18.06 i.e., **72.24 dB**

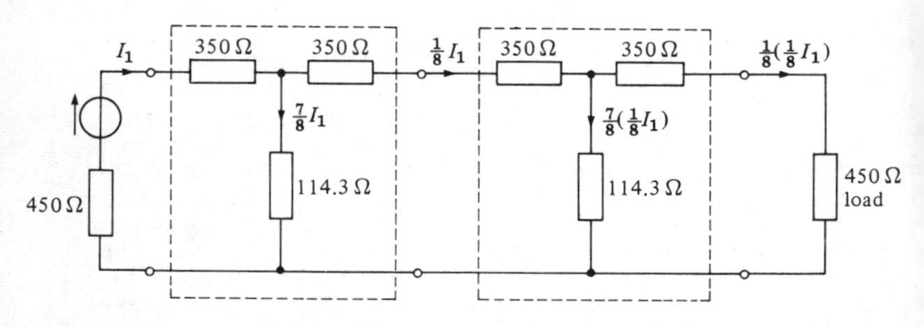

Fig 29

C. FURTHER PROBLEMS ON SIMPLE FILTER AND ATTENUATION CIRCUITS

SHORT ANSWER PROBLEMS

1 What is an electrical 'filter?'
2 Define the term 'cut-off frequency' as applied to a filter.
3 Describe briefly the function of a low-pass filter. Give one application.
4 Sketch (a) ideal (b) practical, attenuation/frequency characteristics for a low-pass filter.
5 Describe briefly the function of a high-pass filter.
6 Sketch (a) ideal (b) practical, attenuation/frequency characteristics for a high-pass filter.
7 Describe briefly the function of a band-pass filter.
8 Sketch (a) ideal (b) practical, attenuation/frequency characteristics for a band-pass filter.
9 Describe briefly the function of a band–stop filter.
10 Sketch (a) ideal (b) practical, attenuation/frequency characteristics for a band-stop filter.
11 What is an attenuator?
12 What is meant by a 'two-port network'? Give an example.
13 Explain briefly the terms (a) iterative impedance
(b) characteristic impedance
14 Derive equations for the characteristic impedance R_o of (a) a symmetrical T-attenuator (b) a symmetrical π-attenuator, in terms of the series resistance R_1 and shunt resistance R_2.
15 What is meant by 'cascading' attenuators?

MULTI-CHOICE PROBLEMS (answers on page 191)

1 A filter designed to pass signals with frequencies between two specified cut-off frequencies is called a
(a) low-pass filter (b) high-pass filter (c) band-pass filter (d) band-stop filter
2 A filter designed to pass signals at frequencies above a specified cut-off frequency is called a
(a) low-pass filter (b) high-pass filter (c) band-pass filter (d) band-stop filter
3 A filter designed to pass signals at frequencies below a specified cut-off frequency is called a
(a) low-pass filter (b) high-pass filter (c) band-pass filter (d) band-stop filter
4 A filter designed to pass signals with all frequencies except those between two specified cut-off frequencies is called a
(a) low-pass filter (b) high-pass filter (c) band-pass filter (d) band-stop filter
5 A T-section symmetrical attenuator pad with series resistance R_A and shunt resistance R_B is to provide a voltage attenuation of 12.04 dB and have a characteristic impedance of 600 Ω. The value of R_A is
(a) 1125 Ω (b) 360 Ω (c) 320 Ω (d) 1000 Ω
6 For the attenuator described in *problem 5*, the value of R_B is
(a) 1125 Ω (b) 360 Ω (c) 320 Ω (d) 1000 Ω
7 A π-section symmetrical attenuator pad with series resistance R_A and shunt resistance R_B is to provide a voltage attenuation of 19.08 dB and have a characteristic impedance of 400 Ω. The value of R_A is
(a) 1777.8 Ω (b) 90 Ω (c) 500 Ω (d) 320 Ω

8 For the attenuator described in *problem 7*, the value of R_B is
(a) 1777.8 Ω (b) 90 Ω (c) 500 Ω (d) 320 Ω

9 The ratio of input to output voltage for a symmetrical T-attenuator is 10. Five identical such attenuators are cascaded. The overall attenuation is
(a) 50 dB (b) 10 dB (c) 100 dB (d) 5 dB

CONVENTIONAL PROBLEMS

1 Determine the characteristic impedances of the T-network attenuator sections shown in *Fig 30*. [(a) 26.46 Ω (b) 244.9 Ω (c) 1.342 kΩ]

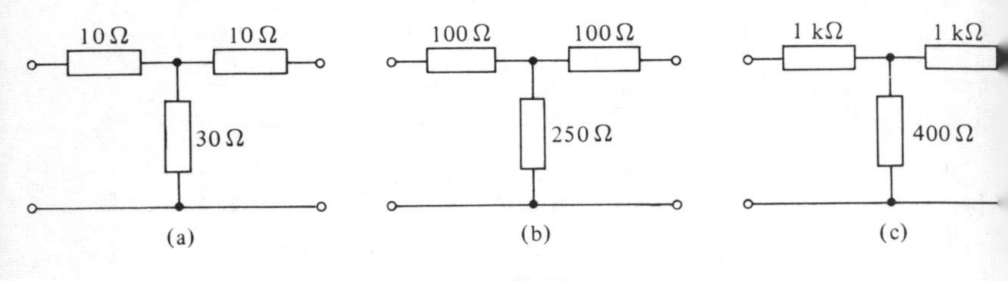

Fig 30

2 Determine the characteristic impedances of the π-network attenuator pads shown in *Fig 31*. [(a) 7.45 Ω (b) 353.6 Ω (c) 189.7 Ω]

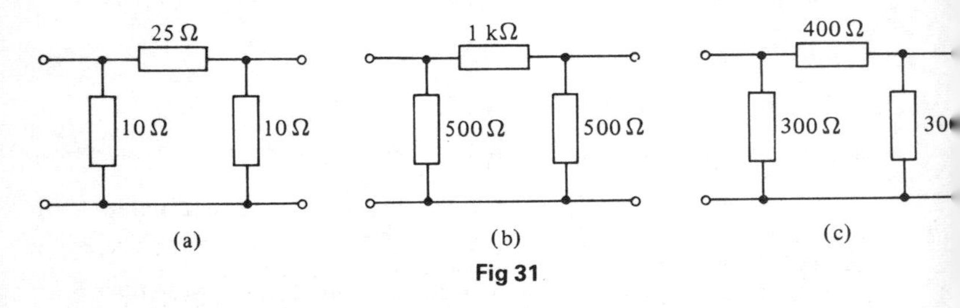

Fig 31

3 A T-section attenuator is to provide 18 dB voltage attenuation per section and is to match a 1.5 kΩ line. Determine the resistance values necessary per section. [$R_1 = 1165\ \Omega$, $R_2 = 384\ \Omega$]

4 A π-section attenuator has a series resistance of 500 Ω and shunt resistances of 2 kΩ. Determine (a) the characteristic impedance, and (b) the attenuation produced by the network. [(a) 667 Ω (b) 6 dB]

5 For each of the attenuator pads shown in *Fig 32* determine (a) the input resistance when the output port is open-circuited, (b) the input resistance when the output port is short-circuited, and (c) the characteristic impedance.
[(i) (a) 50 Ω (b) 42 Ω (c) 45.83 Ω
(ii) (a) 285.7 Ω (b) 240 Ω (c) 261.9 Ω]

6 A television signal received from an aerial through a length of coaxial cable of characteristic impedance 100 Ω has to be attenuated by 15 dB before entering the receiver. If the input impedance of the receiver is also 100 Ω, design a suitable T-attenuator network to give the necessary reduction. [$R_1 = 69.8\ \Omega$, $R_2 = 36.7\ \Omega$]

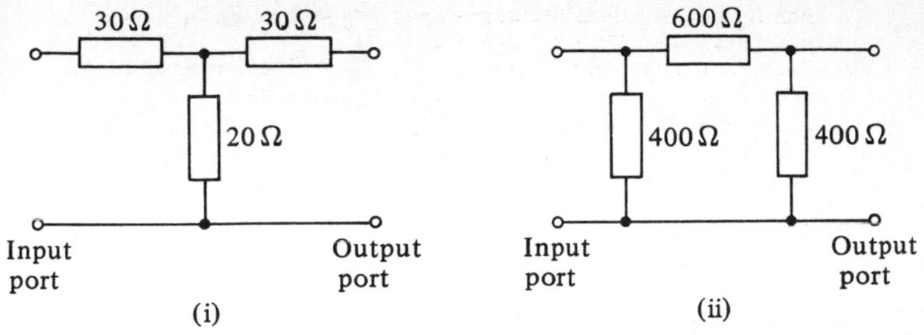

Fig 32

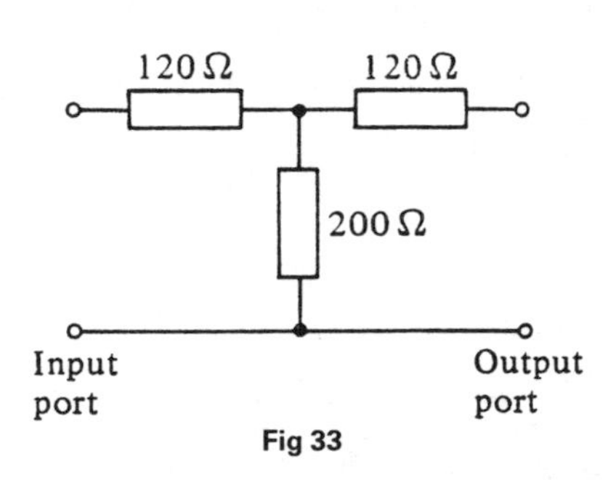

Fig 33

7 Design (a) a T-section symmetrical attenuator pad, and (b) a π-section symmetrical attenuator pad, to provide a voltage attenuation of 15 dB and having a characteristic impedance of 500 Ω.

[(a) $R_1 = 349\ \Omega$, $R_2 = 184\ \Omega$
(b) $R_1 = 1.36\ k\Omega$, $R_2 = 716\ \Omega$]

8 Determine the values of the shunt and series resistances for T-pad attenuators of characteristic impedance 400 Ω to provide the following attenuations: (a) 12 dB (b) 25 dB (c) 36 dB.

[(a) $R_1 = 239.4\ \Omega$, $R_2 = 214.5\ \Omega$
(b) $R_1 = 357.4\ \Omega$, $R_2 = 45.14\ \Omega$
(c) $R_1 = 387.5\ \Omega$, $R_2 = 12.68\ \Omega$]

9 Design a π-section symmetrical attenuator network to provide a voltage attenuation of 24 dB and having a characteristic impedance of 600 Ω.

[$R_1 = 4.736\ k\Omega$, $R_2 = 680.8\ \Omega$]

10 Explain what is meant by 'the characteristic impedance of an attenuator section'. Determine the values of the shunt and series resistance for π-pad attenuator sections of characteristic impedance 600 Ω to give the following attenuations:
(a) 8 dB (b) 20 dB (c) 32 dB.

[(a) $R_1 = 634.1\ \Omega$, $R_2 = 1393.7\ \Omega$
(b) $R_1 = 2.97\ k\Omega$, $R_2 = 733.3\ \Omega$
(c) $R_1 = 11.94\ k\Omega$, $R_2 = 630.9\ \Omega$]

11 A battery of emf E and negligible internal resistance is connected across the input terminals of the T-network shown in *Fig 33*. Determine, in terms of E, the current drawn from the battery when (a) the output terminals are open-circuited, (b) the output terminals are short-circuited, (c) the network is

correctly terminated. (d) For the last case, determine the attenuation of the network in decibels.

[(a) $\frac{E}{320}$ A (b) $\frac{E}{195}$ A (c) $\frac{E}{249.8}$ A
(d) 9.09 dB]

12 A d.c. generator has an internal resistance of 600 Ω and supplies a 600 Ω load. Design a symmetrical (a) T-network and (b) π-network attenuator pad, having a characteristic impedance of 600 Ω which when connected between the generator and load will reduce the load current to $\frac{1}{4}$ its initial value.

[(a) R_1 = 360 Ω, R_2 = 320 Ω
(b) R_1 = 1125 Ω, R_2 = 1000 Ω]

13 The input to an attenuator is 24 V and the output is 4 V. Determine the attenuation in decibels. If five such identical attenuators are cascaded, determine the overall attenuation. [15.56 dB; 77.80 dB]

14 Four identical attenuator sections are connected in cascade. The overall attenuation is 60 dB. The input to the first section is 50 mV. Determine (a) the attenuation of each section, (b) the output of the final stage, and (c) the output of the second stage. [(a) 15 dB (b) 50 μV (c) 1.58 mV]

15 A d.c. generator has an internal resistance of 300 Ω and supplies a 300 Ω load.
(a) Design a symmetrical T-network attenuator pad having a characteristic impedance of 300 Ω which, when connected between the generator and the load, will reduce the load current to $\frac{1}{3}$ its initial value.
(b) If two such networks as in (a) were connected in series between the generator and the load, what fraction of the initial current would the load take?
(c) Determine the fraction of the initial current that the load would take if six such networks were cascaded between the generator and the load.
(d) Determine the attenuation in decibels provided by five such identical stages as in (a).

[(a) R_1 = 150 Ω, R_2 = 225 Ω
(b) 1/9 (c) 1/729 (d) 47.71 dB]

Answers to multi-choice problems

Chapter 1 (page 30): 1 (b); 2 (c); (3) (a); 4 (d); 5 (c); 6 (a); 7 (c); 8 (c); 9 (b); 10 (d); 11 (b).
Chapter 2 (page 53): 1 (b); 2 (a); 3 (b); 4 (b); 5 (a); 6 (c); 7 (a); 8 (d); 9 (b); 10 (d); 11 (c); 12 (b).
Chapter 3 (page 73): 1 (d); 2 (g); 3 (i); 4 (s); 5 (h); 6 (b); 7 (k); 8 (l); 9 (a); 10 (d), (g), (i) and (l); 11 (b); 12 (d).
Chapter 4 (page 94): 1 (g); 2 (c); 3 (a); 4 (a); 5 (f); 6 (a); 7 (g); 8 (l); 9 (l); 10 (d); 11 (f); 12 (j).
Chapter 5 (page 113): 1 (c); 2 (b); 3 (b); 4 (g); 5 (g); 6 (e); 7 (l); 8 (c); 9 (a); 10 (d); 11 (g); 12 (b); 13 (c); 14 (j); 15 (h).
Chapter 6 (page 128): 1 (b); 2 (e); 3 (e); 4 (c); 5 (c); 6 (a); 7 (d); 8 (f); 9 (b); 10 (c); 11 (b); 12 (a); 13 (b); 14 (a); 15 (d).
Chapter 7 (page 137): 1 (c); 2 (b); 3 (d); 4 (a); 5 (b); 6 (d); 7 (b); 8 (c); 9 (f); 10 (j).
Chapter 8 (page 148): 1 (d); 2 (a); 3 (a); 4 (b); 5 (d).
Chapter 9 (page 163): 1 (b); 2 (d); 3 (a); 4 (d); 5 (c); 6 (g); 7 (c); 8 (b); 9 (p); 10 (d); 11 (o); 12 (n).
Chapter 10 (page 187): 1 (c); 2 (b); 3 (a); 4 (d); 5 (b); 6 (c); 7 (d); 8 (a); 9 (c).

Index